100 DAYS OF TIMED TESTS

ADDITION AND SUBTRACTION WORKBOOK

DIGITS 0-20
MATH DRILLS
AND REPRODUCIBLE
PRACTICE PROBLEMS
WORKBOOKS
KINDERGARTEN

LUCAS VINCENT

100 Days of Timed Tests Addition And Subtraction Workbook

First Published in 2023 by Lucas Vincent

Contents

DAY 01
ADDING DIGITS 0–6

SCORE: | TIME: :

0 + 1	0 + 9	0 + 9	0 + 7
0 + 9	0 + 8	0 + 4	0 + 6
0 + 10	0 + 9	0 + 8	0 + 7
0 + 6	0 + 3	0 + 8	0 + 8
0 + 6	0 + 10	0 + 1	0 + 9
0 + 4	0 + 7	0 + 8	0 + 6
0 + 2	0 + 10	0 + 6	0 + 9
0 + 10	0 + 4	0 + 4	0 + 6
0 + 8	0 + 9	0 + 3	0 + 10
0 + 7	0 + 8	0 + 2	0 + 8

DAY 02

ADDING DIGITS
0–6

SCORE: | TIME: :

1 + 0	1 + 8	1 + 9	1 + 7
1 + 1	1 + 6	1 + 3	1 + 8
1 + 9	1 + 7	1 + 6	1 + 4
1 + 8	1 + 10	1 + 5	1 + 3
1 + 9	1 + 9	1 + 9	1 + 4
1 + 10	1 + 0	1 + 10	1 + 0
1 + 10	1 + 4	1 + 7	1 + 5
1 + 5	1 + 8	1 + 9	1 + 7
1 + 3	1 + 4	1 + 7	1 + 9
1 + 2	1 + 5	1 + 0	1 + 2

DAY 03

ADDING DIGITS 0–6

SCORE: | TIME: :

2 + 7	2 + 0	2 + 4	2 + 5
2 + 1	2 + 3	2 + 10	2 + 3
2 + 4	2 + 0	2 + 7	2 + 0
2 + 10	2 + 3	2 + 7	2 + 1
2 + 0	2 + 3	2 + 2	2 + 3
2 + 0	2 + 1	2 + 3	2 + 4
2 + 8	2 + 2	2 + 2	2 + 1
2 + 4	2 + 8	2 + 9	2 + 6
2 + 7	2 + 10	2 + 6	2 + 4
2 + 4	2 + 7	2 + 2	2 + 7

DAY 04
ADDING DIGITS 0–6

SCORE: | TIME: :

3 + 2	3 + 10	3 + 7	3 + 8
3 + 1	3 + 7	3 + 2	3 + 10
3 + 5	3 + 10	3 + 6	3 + 8
3 + 2	3 + 8	3 + 5	3 + 10
3 + 3	3 + 6	3 + 3	3 + 6
3 + 6	3 + 9	3 + 5	3 + 4
3 + 6	3 + 1	3 + 0	3 + 6
3 + 0	3 + 6	3 + 7	3 + 10
3 + 6	3 + 0	3 + 1	3 + 2
3 + 8	3 + 9	3 + 6	3 + 2

DAY 05

ADDING DIGITS 0–6

SCORE: | TIME: :

4 + 9	4 + 0	4 + 7	4 + 9
4 + 3	4 + 0	4 + 2	4 + 3
4 + 3	4 + 6	4 + 7	4 + 4
4 + 1	4 + 1	4 + 4	4 + 4
4 + 0	4 + 8	4 + 8	4 + 9
4 + 3	4 + 3	4 + 0	4 + 9
4 + 6	4 + 8	4 + 7	4 + 6
4 + 8	4 + 1	4 + 9	4 + 4
4 + 2	4 + 0	4 + 8	4 + 10
4 + 8	4 + 1	4 + 9	4 + 6

DAY 06

ADDING DIGITS 0–6

SCORE: | TIME: :

5 + 7	5 + 5	5 + 7	5 + 4
5 + 3	5 + 0	5 + 7	5 + 0
5 + 0	5 + 5	5 + 5	5 + 5
5 + 7	5 + 4	5 + 3	5 + 9
5 + 3	5 + 3	5 + 9	5 + 0
5 + 10	5 + 9	5 + 2	5 + 6
5 + 4	5 + 4	5 + 8	5 + 7
5 + 3	5 + 6	5 + 4	5 + 9
5 + 9	5 + 10	5 + 1	5 + 3
5 + 1	5 + 2	5 + 9	5 + 3

DAY 07

ADDING DIGITS

0–6

SCORE: | TIME: :

6 + 10	6 + 3	6 + 10	6 + 0
6 + 4	6 + 6	6 + 1	6 + 5
6 + 4	6 + 2	6 + 2	6 + 4
6 + 6	6 + 1	6 + 5	6 + 0
6 + 8	6 + 6	6 + 2	6 + 3
6 + 2	6 + 9	6 + 10	6 + 5
6 + 8	6 + 6	6 + 3	6 + 10
6 + 1	6 + 4	6 + 3	6 + 4
6 + 10	6 + 5	6 + 5	6 + 6
6 + 2	6 + 3	6 + 9	6 + 5

DAY 08

ADDING DIGITS

0–5

SCORE: | TIME: :

2 + 1	1 + 0	4 + 2	3 + 5
2 + 3	2 + 2	3 + 0	2 + 3
5 + 0	5 + 0	4 + 3	2 + 2
5 + 2	2 + 1	1 + 5	5 + 1
4 + 1	5 + 0	4 + 1	3 + 2
4 + 1	2 + 4	3 + 2	1 + 5
1 + 3	1 + 3	2 + 4	1 + 5
5 + 0	3 + 2	3 + 4	2 + 5
2 + 5	5 + 1	4 + 4	2 + 1
1 + 4	3 + 1	4 + 3	5 + 5

DAY 09

ADDING DIGITS
0–5

SCORE: | TIME: :

$4 + 1$	$3 + 3$	$1 + 0$	$4 + 1$
$3 + 0$	$3 + 4$	$1 + 4$	$3 + 5$
$1 + 5$	$1 + 5$	$5 + 2$	$3 + 5$
$1 + 4$	$1 + 3$	$4 + 4$	$3 + 0$
$1 + 3$	$4 + 5$	$1 + 2$	$5 + 2$
$5 + 2$	$3 + 5$	$1 + 2$	$2 + 5$
$5 + 0$	$1 + 0$	$5 + 4$	$2 + 5$
$3 + 5$	$3 + 3$	$3 + 5$	$1 + 2$
$3 + 1$	$5 + 1$	$3 + 0$	$3 + 2$
$5 + 1$	$1 + 5$	$3 + 4$	$1 + 3$

DAY 10

ADDING DIGITS

0–5

SCORE: | TIME: :

5 + 4	2 + 3	2 + 4	4 + 1
4 + 3	5 + 4	3 + 1	2 + 3
5 + 2	4 + 1	5 + 2	2 + 3
2 + 4	4 + 0	2 + 0	4 + 4
2 + 5	3 + 0	2 + 2	4 + 5
5 + 4	1 + 2	1 + 3	2 + 5
4 + 3	5 + 5	1 + 0	1 + 5
1 + 2	5 + 4	3 + 1	1 + 3
1 + 4	4 + 2	4 + 2	2 + 4
1 + 1	2 + 2	3 + 4	4 + 5

DAY 11

ADDING DIGITS

0–5

SCORE: | TIME: :

3 + 1	4 + 5	5 + 5	5 + 0
3 + 1	3 + 1	4 + 5	1 + 0
5 + 0	2 + 4	3 + 5	2 + 4
4 + 4	1 + 1	5 + 2	3 + 1
1 + 2	2 + 3	3 + 4	2 + 5
4 + 1	2 + 5	3 + 1	5 + 3
5 + 4	3 + 0	5 + 4	4 + 2
4 + 1	3 + 2	5 + 1	5 + 2
2 + 0	4 + 1	5 + 3	3 + 0
2 + 1	4 + 4	2 + 0	3 + 0

DAY 12
ADDING DIGITS
0–5

SCORE: | TIME: :

1 + 2	4 + 2	3 + 4	2 + 3
4 + 5	1 + 5	3 + 3	1 + 5
1 + 5	3 + 0	2 + 1	5 + 4
5 + 1	2 + 5	4 + 1	4 + 1
4 + 2	5 + 3	4 + 0	1 + 0
3 + 2	2 + 3	3 + 1	4 + 5
1 + 5	1 + 2	1 + 4	4 + 5
1 + 4	2 + 0	5 + 1	3 + 2
1 + 4	5 + 2	1 + 4	2 + 3
1 + 5	4 + 0	5 + 0	5 + 2

DAY 13

ADDING DIGITS

0–5

SCORE: | TIME: :

4 + 2	5 + 1	4 + 4	2 + 1
1 + 3	5 + 2	4 + 4	4 + 3
1 + 2	1 + 1	5 + 0	5 + 4
3 + 3	4 + 0	3 + 4	2 + 0
3 + 4	4 + 0	3 + 4	3 + 0
3 + 3	1 + 1	2 + 0	2 + 1
2 + 5	5 + 2	2 + 4	2 + 5
4 + 2	5 + 3	3 + 0	5 + 1
1 + 0	1 + 3	4 + 3	1 + 5
5 + 2	1 + 3	2 + 2	1 + 0

DAY 14
ADDING DIGITS
0–5

SCORE: | TIME: :

3 + 4	3 + 5	4 + 3	1 + 4
2 + 5	4 + 0	2 + 4	4 + 1
1 + 5	2 + 3	5 + 0	2 + 4
2 + 5	2 + 0	1 + 0	1 + 1
1 + 4	4 + 1	4 + 2	3 + 3
5 + 2	3 + 1	1 + 4	4 + 0
4 + 2	4 + 1	2 + 0	4 + 1
4 + 4	5 + 5	1 + 1	1 + 5
5 + 0	4 + 1	4 + 2	1 + 0
2 + 3	5 + 3	5 + 3	4 + 1

DAY 15

ADDING DIGITS 0–5

SCORE: | TIME: :

$2 + 5 =$	$2 + 4 =$	$5 + 0 =$	$2 + 0 =$
$3 + 3 =$	$1 + 0 =$	$1 + 2 =$	$4 + 4 =$
$3 + 5 =$	$1 + 5 =$	$3 + 1 =$	$5 + 1 =$
$4 + 2 =$	$3 + 1 =$	$5 + 5 =$	$1 + 3 =$
$1 + 4 =$	$3 + 1 =$	$4 + 0 =$	$2 + 5 =$
$5 + 2 =$	$2 + 2 =$	$3 + 3 =$	$1 + 4 =$
$4 + 1 =$	$5 + 1 =$	$2 + 0 =$	$4 + 0 =$
$5 + 1 =$	$2 + 0 =$	$4 + 2 =$	$2 + 1 =$
$4 + 4 =$	$3 + 4 =$	$5 + 5 =$	$2 + 1 =$
$5 + 5 =$	$1 + 5 =$	$1 + 2 =$	$3 + 2 =$

DAY 16

ADDING DIGITS

0–5

SCORE: | TIME: :

1 + 2	5 + 5	5 + 3	2 + 0
5 + 0	3 + 0	4 + 0	1 + 4
4 + 5	3 + 1	1 + 1	4 + 0
4 + 1	4 + 0	4 + 3	4 + 5
4 + 1	4 + 0	3 + 5	3 + 3
4 + 3	4 + 2	3 + 4	3 + 0
1 + 1	1 + 0	2 + 0	2 + 0
1 + 2	2 + 5	3 + 0	5 + 2
5 + 5	5 + 1	4 + 0	2 + 4
2 + 1	2 + 0	5 + 3	4 + 0

DAY 17
ADDING DIGITS
0–10

SCORE: | TIME: :

9 + 2	3 + 2	4 + 4	5 + 0
2 + 0	7 + 3	1 + 2	8 + 1
5 + 3	5 + 4	6 + 3	5 + 4
4 + 3	4 + 5	1 + 2	3 + 4
2 + 3	7 + 2	6 + 2	9 + 5
6 + 3	5 + 5	3 + 2	4 + 1
5 + 5	2 + 4	1 + 1	5 + 1
5 + 2	8 + 5	7 + 5	5 + 5
4 + 5	9 + 2	7 + 4	9 + 5
8 + 2	5 + 2	9 + 0	7 + 4

DAY 18

ADDING DIGITS 0–10

SCORE: | TIME: :

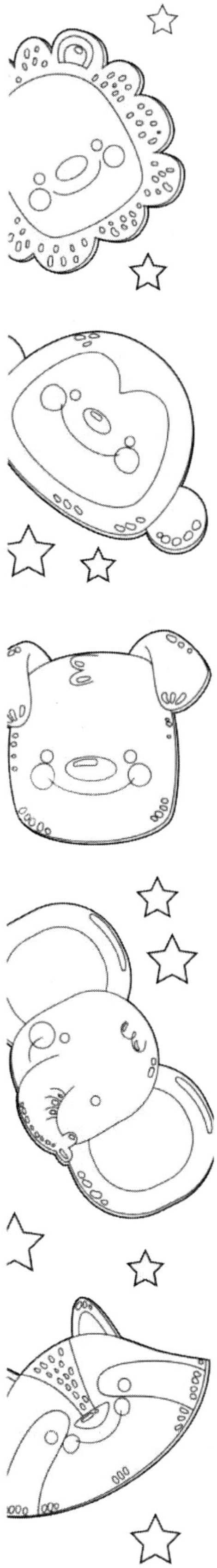

2 + 4 =	6 + 0 =	5 + 4 =	7 + 3 =
4 + 0 =	7 + 0 =	4 + 2 =	5 + 3 =
9 + 2 =	2 + 1 =	1 + 4 =	7 + 0 =
5 + 5 =	5 + 3 =	8 + 2 =	3 + 5 =
1 + 4 =	2 + 0 =	1 + 4 =	9 + 3 =
7 + 3 =	9 + 1 =	7 + 4 =	6 + 2 =
7 + 0 =	9 + 1 =	7 + 3 =	5 + 5 =
7 + 0 =	2 + 5 =	1 + 3 =	3 + 0 =
5 + 0 =	7 + 2 =	7 + 4 =	1 + 1 =
7 + 1 =	2 + 5 =	8 + 2 =	7 + 4 =

DAY 19

ADDING DIGITS 0–10

SCORE: | TIME: :

9 + 4	9 + 4	5 + 0	8 + 2
9 + 2	2 + 4	4 + 1	8 + 0
3 + 3	4 + 4	6 + 4	1 + 5
5 + 3	5 + 2	1 + 1	5 + 1
7 + 4	5 + 1	7 + 3	1 + 2
4 + 5	2 + 4	2 + 0	1 + 2
1 + 5	6 + 1	4 + 2	6 + 4
4 + 4	9 + 4	5 + 4	7 + 1
6 + 2	7 + 5	3 + 5	4 + 1
2 + 4	3 + 1	2 + 2	5 + 5

DAY 20

ADDING DIGITS

0–10

SCORE: | TIME: :

5 + 0	7 + 1	3 + 5	6 + 1
8 + 1	3 + 2	7 + 5	8 + 1
6 + 3	6 + 2	9 + 5	3 + 5
8 + 0	8 + 3	4 + 0	9 + 0
8 + 1	2 + 3	3 + 0	1 + 5
4 + 2	5 + 4	2 + 5	7 + 5
1 + 0	2 + 2	3 + 1	1 + 3
3 + 4	7 + 1	9 + 4	3 + 1
5 + 0	7 + 0	8 + 3	1 + 1
4 + 4	2 + 0	5 + 5	4 + 0

DAY 21

ADDING DIGITS

0–10

SCORE: | TIME: :

9 + 3	8 + 1	1 + 3	6 + 6
4 + 2	5 + 5	8 + 6	9 + 0
2 + 5	8 + 2	4 + 4	7 + 4
8 + 0	2 + 4	9 + 0	2 + 2
6 + 0	6 + 6	4 + 6	3 + 0
9 + 2	4 + 0	6 + 3	2 + 5
9 + 3	9 + 0	4 + 3	7 + 3
1 + 5	1 + 4	3 + 1	2 + 0
2 + 4	3 + 5	8 + 3	7 + 0
6 + 0	2 + 3	6 + 4	1 + 3

DAY 22

ADDING DIGITS

0–10

SCORE: | TIME: :

2 + 6	5 + 6	9 + 0	3 + 2
5 + 0	4 + 4	3 + 2	1 + 1
6 + 4	3 + 6	4 + 1	2 + 6
7 + 5	9 + 2	5 + 1	5 + 3
8 + 6	2 + 5	6 + 4	7 + 4
8 + 6	3 + 2	2 + 5	2 + 3
3 + 6	5 + 3	2 + 1	8 + 4
1 + 0	1 + 4	1 + 4	5 + 6
6 + 3	4 + 3	2 + 0	6 + 4
9 + 5	2 + 1	6 + 4	1 + 6

DAY 23

ADDING DIGITS

0–10

SCORE: | TIME: :

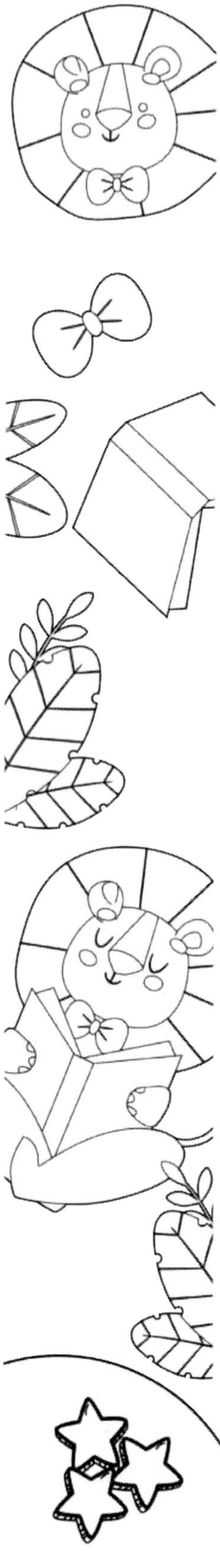

2 + 3	9 + 5	7 + 3	6 + 0
9 + 0	8 + 0	5 + 0	7 + 1
3 + 3	2 + 5	8 + 3	9 + 3
8 + 0	5 + 2	4 + 6	3 + 4
3 + 0	8 + 4	8 + 3	1 + 5
3 + 2	5 + 1	4 + 0	7 + 4
5 + 5	2 + 6	7 + 0	5 + 0
3 + 3	2 + 0	5 + 2	8 + 1
6 + 5	7 + 1	8 + 3	8 + 6
2 + 5	1 + 2	8 + 1	4 + 5

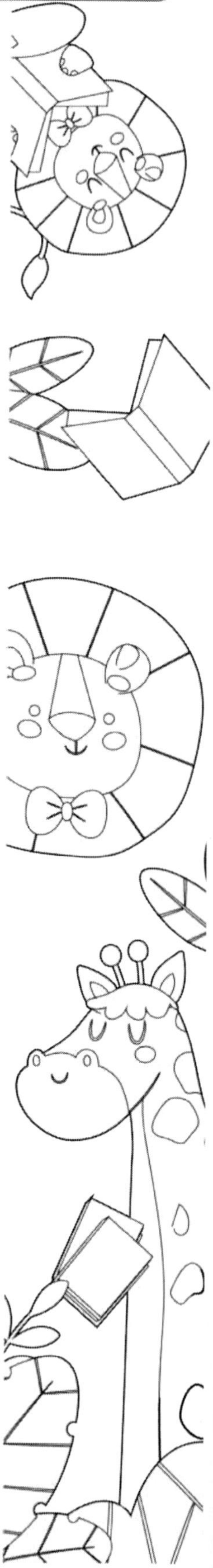

DAY 24

ADDING DIGITS

0–10

SCORE: | TIME: :

5 + 2	1 + 5	6 + 6	2 + 5
3 + 1	9 + 3	8 + 1	5 + 5
4 + 1	1 + 1	1 + 5	2 + 5
8 + 7	3 + 4	6 + 4	5 + 0
5 + 2	9 + 5	1 + 5	4 + 0
7 + 4	1 + 1	3 + 0	5 + 0
2 + 2	6 + 5	4 + 2	3 + 3
6 + 0	3 + 3	5 + 7	9 + 2
8 + 4	3 + 5	8 + 3	8 + 6
3 + 1	4 + 5	2 + 3	3 + 0

DAY 25

ADDING DIGITS 0–10

SCORE: | TIME: :

3 + 0	7 + 5	9 + 4	9 + 2
7 + 6	4 + 2	3 + 3	4 + 6
3 + 7	6 + 7	5 + 3	8 + 0
3 + 0	5 + 3	2 + 4	8 + 7
4 + 0	3 + 4	3 + 7	8 + 6
5 + 1	1 + 1	6 + 6	8 + 2
6 + 1	6 + 6	9 + 6	6 + 5
5 + 6	3 + 5	4 + 3	1 + 7
3 + 0	2 + 0	3 + 7	5 + 6
3 + 5	5 + 5	9 + 4	3 + 3

DAY 26

ADDING DIGITS

0–10

SCORE: | TIME: :

7 + 3 = ___	3 + 6 = ___	2 + 6 = ___	1 + 4 = ___
9 + 0 = ___	7 + 3 = ___	6 + 0 = ___	6 + 1 = ___
9 + 3 = ___	5 + 0 = ___	4 + 2 = ___	3 + 4 = ___
5 + 3 = ___	6 + 2 = ___	2 + 6 = ___	2 + 1 = ___
2 + 1 = ___	1 + 5 = ___	7 + 5 = ___	4 + 0 = ___
4 + 1 = ___	9 + 2 = ___	9 + 0 = ___	1 + 0 = ___
3 + 7 = ___	7 + 3 = ___	7 + 6 = ___	9 + 6 = ___
9 + 6 = ___	8 + 5 = ___	3 + 2 = ___	6 + 0 = ___
6 + 5 = ___	5 + 2 = ___	3 + 4 = ___	8 + 1 = ___
8 + 6 = ___	8 + 0 = ___	5 + 0 = ___	1 + 1 = ___

DAY 27

ADDING DIGITS

0–10

SCORE: | TIME: :

1 + 2 =	3 + 3 =	6 + 7 =	4 + 3 =
2 + 4 =	8 + 2 =	6 + 6 =	8 + 7 =
3 + 5 =	5 + 6 =	9 + 3 =	2 + 5 =
7 + 7 =	9 + 0 =	7 + 2 =	4 + 5 =
1 + 6 =	2 + 4 =	7 + 8 =	8 + 0 =
6 + 7 =	3 + 5 =	8 + 6 =	8 + 0 =
9 + 8 =	6 + 1 =	1 + 6 =	4 + 4 =
8 + 5 =	1 + 6 =	7 + 3 =	8 + 7 =
2 + 7 =	6 + 1 =	1 + 7 =	5 + 2 =
9 + 1 =	6 + 6 =	8 + 4 =	6 + 7 =

DAY 28

ADDING DIGITS

0–10

SCORE: | TIME: :

4 + 3	5 + 3	5 + 8	4 + 7
3 + 3	7 + 0	8 + 8	4 + 1
3 + 6	9 + 2	8 + 1	7 + 0
6 + 5	2 + 3	4 + 0	5 + 4
1 + 3	8 + 2	8 + 8	9 + 7
3 + 5	1 + 7	8 + 4	9 + 7
7 + 2	4 + 2	7 + 6	9 + 8
5 + 4	9 + 4	8 + 7	2 + 5
4 + 2	8 + 5	7 + 4	7 + 6
9 + 5	1 + 6	1 + 7	5 + 5

DAY 29

ADDING DIGITS

0–10

SCORE: | TIME: :

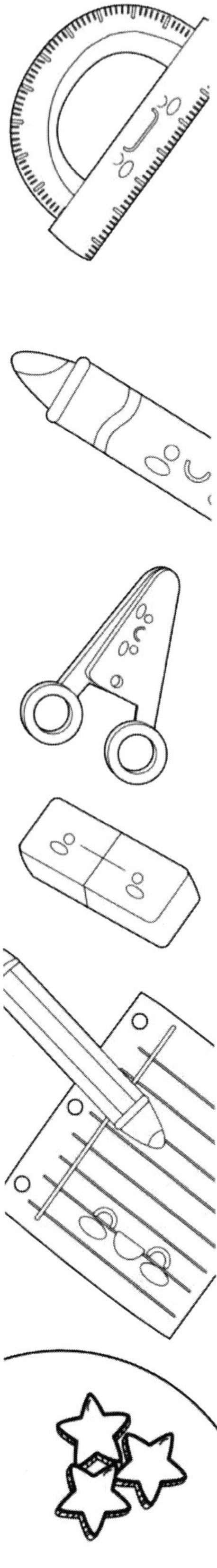

4 + 6	4 + 6	6 + 5	7 + 4
7 + 1	8 + 7	3 + 2	2 + 1
7 + 4	8 + 3	1 + 0	1 + 6
9 + 2	4 + 0	4 + 2	8 + 8
2 + 0	4 + 1	7 + 1	9 + 2
9 + 1	4 + 3	2 + 2	6 + 7
4 + 6	2 + 4	7 + 2	7 + 6
7 + 0	1 + 7	3 + 5	6 + 0
2 + 0	8 + 4	5 + 4	3 + 5
1 + 2	6 + 3	2 + 2	5 + 4

DAY 30

ADDING DIGITS

0–10

SCORE: | TIME: :

6 + 4	9 + 4	2 + 2	4 + 5
8 + 6	5 + 4	5 + 2	3 + 1
9 + 0	2 + 7	1 + 5	4 + 3
3 + 5	7 + 1	2 + 4	2 + 2
8 + 3	4 + 4	7 + 4	9 + 3
1 + 8	7 + 0	5 + 3	1 + 7
6 + 6	5 + 6	6 + 0	5 + 6
2 + 0	4 + 5	6 + 5	7 + 3
8 + 3	1 + 1	9 + 3	3 + 6
9 + 0	5 + 3	1 + 4	7 + 3

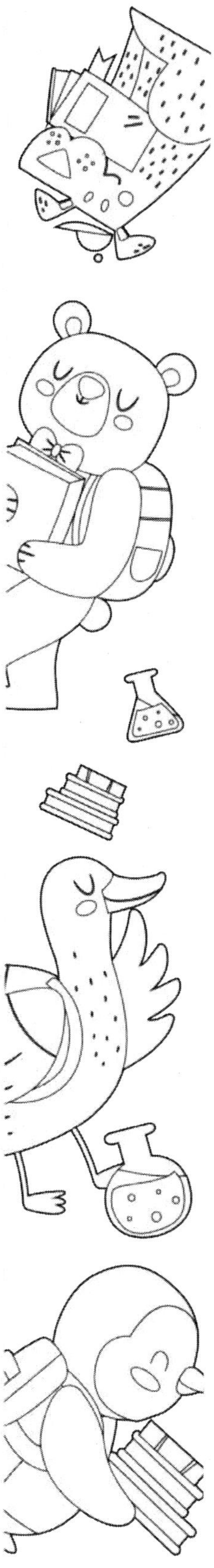

DAY 31

ADDING DIGITS

0–10

SCORE: | TIME: :

6 + 5	6 + 3	2 + 8	6 + 4
6 + 9	4 + 2	9 + 6	2 + 4
2 + 6	3 + 2	5 + 5	3 + 2
2 + 7	7 + 9	9 + 5	2 + 1
6 + 0	3 + 4	7 + 0	2 + 3
6 + 4	1 + 7	6 + 2	6 + 1
2 + 9	1 + 8	1 + 2	2 + 2
4 + 5	2 + 5	8 + 6	4 + 0
6 + 5	3 + 8	8 + 9	8 + 6
8 + 3	7 + 0	1 + 4	4 + 0

DAY 32

ADDING DIGITS

0–10

SCORE: | TIME: :

2 + 7	6 + 1	7 + 8	5 + 1
5 + 7	8 + 3	6 + 6	1 + 7
3 + 2	3 + 3	1 + 3	9 + 8
6 + 6	2 + 1	6 + 8	6 + 3
9 + 1	1 + 2	2 + 7	5 + 7
5 + 2	7 + 5	7 + 3	9 + 4
2 + 2	3 + 0	4 + 5	2 + 7
2 + 0	2 + 3	7 + 0	6 + 9
3 + 6	4 + 3	4 + 9	4 + 0
3 + 2	4 + 7	1 + 4	8 + 5

DAY 33

ADDING DIGITS 0–10

SCORE: | TIME: :

9 + 1	9 + 3	9 + 7	9 + 7
4 + 1	5 + 5	5 + 4	4 + 3
1 + 5	4 + 1	8 + 9	3 + 5
6 + 6	3 + 9	8 + 6	2 + 1
9 + 0	1 + 0	8 + 8	4 + 1
9 + 3	9 + 1	9 + 7	3 + 4
9 + 2	2 + 9	5 + 8	2 + 5
4 + 7	1 + 3	9 + 9	6 + 4
4 + 6	1 + 5	5 + 2	9 + 5
8 + 1	4 + 7	8 + 9	7 + 0

DAY 34

ADDING DIGITS

0–10

SCORE: | TIME: :

1 + 5	8 + 6	3 + 5	4 + 3
8 + 3	8 + 8	9 + 3	5 + 0
1 + 0	3 + 2	9 + 5	4 + 4
8 + 0	7 + 4	4 + 9	8 + 7
4 + 5	9 + 6	7 + 5	6 + 9
5 + 2	4 + 9	2 + 6	7 + 6
9 + 0	6 + 8	5 + 7	5 + 9
8 + 9	4 + 4	7 + 9	1 + 3
7 + 5	2 + 7	7 + 3	2 + 5
6 + 8	9 + 2	5 + 7	5 + 2

DAY 35

ADDING DIGITS 0–10

SCORE: | TIME: :

3 + 0	5 + 2	5 + 10	5 + 10
2 + 7	9 + 3	6 + 10	7 + 1
7 + 1	1 + 1	8 + 5	6 + 10
3 + 6	5 + 0	6 + 6	5 + 3
2 + 6	6 + 10	4 + 9	8 + 5
3 + 6	6 + 0	7 + 1	7 + 4
1 + 1	3 + 8	1 + 2	3 + 10
4 + 7	4 + 6	6 + 6	1 + 10
6 + 10	1 + 1	8 + 5	1 + 1
5 + 7	1 + 10	5 + 1	4 + 7

DAY 36

ADDING DIGITS

0–10

SCORE: | **TIME:** :

7 + 6	3 + 3	9 + 1	7 + 7
4 + 7	2 + 1	8 + 0	3 + 9
8 + 7	9 + 2	2 + 9	9 + 5
6 + 9	4 + 1	6 + 1	3 + 8
2 + 1	8 + 7	4 + 5	5 + 3
5 + 0	2 + 7	4 + 10	1 + 6
6 + 6	8 + 5	1 + 6	5 + 1
6 + 6	9 + 3	3 + 5	3 + 7
9 + 2	4 + 10	5 + 4	2 + 8
8 + 6	6 + 6	4 + 1	2 + 4

DAY 37

ADDING DIGITS

0–10

SCORE: | TIME: :

$5 + 7$	$4 + 6$	$6 + 4$	$7 + 1$
$4 + 3$	$9 + 0$	$6 + 4$	$5 + 2$
$4 + 1$	$1 + 6$	$8 + 1$	$1 + 5$
$2 + 9$	$1 + 0$	$5 + 10$	$3 + 10$
$8 + 8$	$2 + 7$	$1 + 9$	$9 + 2$
$8 + 5$	$4 + 0$	$3 + 2$	$9 + 0$
$2 + 8$	$9 + 6$	$4 + 7$	$2 + 7$
$3 + 1$	$3 + 9$	$8 + 3$	$9 + 1$
$4 + 3$	$9 + 4$	$4 + 5$	$8 + 0$
$6 + 10$	$7 + 8$	$4 + 10$	$7 + 4$

DAY 38

ADDING DIGITS

0–10

SCORE: | TIME: :

4 + 10	7 + 1	3 + 5	2 + 6
1 + 1	9 + 10	9 + 1	5 + 10
1 + 3	8 + 10	8 + 2	6 + 10
5 + 1	4 + 10	8 + 4	5 + 1
7 + 3	3 + 10	2 + 4	3 + 1
7 + 5	1 + 8	6 + 4	2 + 6
5 + 4	4 + 4	2 + 4	2 + 0
2 + 7	6 + 4	2 + 10	4 + 9
8 + 3	5 + 0	4 + 3	8 + 4
4 + 6	4 + 2	8 + 4	6 + 2

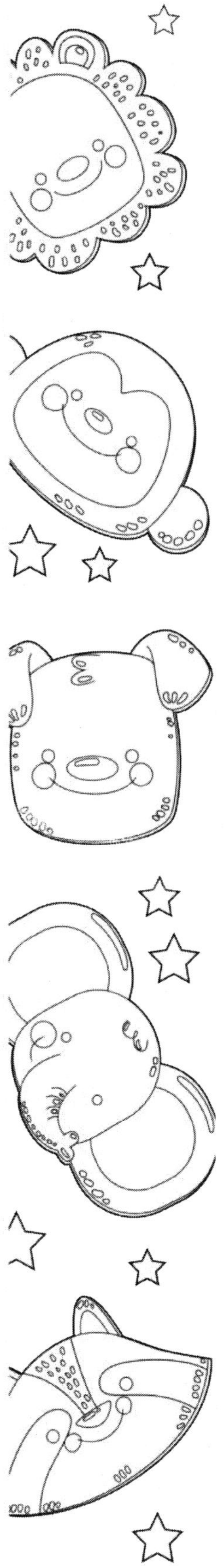

DAY 39
ADDING DIGITS 0–10

SCORE: TIME: :

2 + 0	8 + 7	8 + 7	1 + 9
7 + 1	5 + 1	1 + 9	8 + 3
5 + 8	1 + 6	8 + 9	2 + 8
1 + 8	7 + 7	6 + 10	8 + 5
4 + 3	2 + 5	5 + 0	3 + 4
1 + 6	7 + 9	8 + 7	8 + 9
8 + 0	2 + 0	9 + 5	6 + 10
6 + 4	3 + 5	8 + 7	2 + 5
1 + 2	9 + 5	1 + 8	8 + 7
9 + 1	2 + 1	8 + 10	6 + 1

DAY 40

ADDING DIGITS 0–10

SCORE: | TIME: :

9 + 1	5 + 6	4 + 5	6 + 4
3 + 1	3 + 7	2 + 5	6 + 10
5 + 0	7 + 0	8 + 2	3 + 1
7 + 10	5 + 5	8 + 7	5 + 7
2 + 8	9 + 5	8 + 0	2 + 7
5 + 0	3 + 5	9 + 3	1 + 8
4 + 2	6 + 6	7 + 4	4 + 4
8 + 9	4 + 5	8 + 6	3 + 10
3 + 2	8 + 8	4 + 4	8 + 9
6 + 4	9 + 8	5 + 4	6 + 3

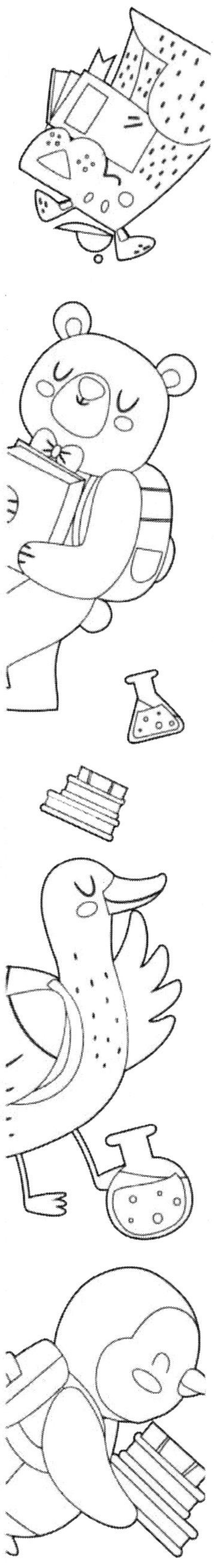

DAY 41

SUBTRACTING DIGITS

0–10

SCORE: | TIME: :

7 - 2	7 - 5	1 - 1	3 - 1
5 - 5	9 - 8	6 - 3	9 - 7
6 - 2	4 - 4	9 - 0	8 - 6
5 - 4	5 - 0	7 - 3	4 - 1
7 - 7	6 - 6	2 - 1	6 - 5
7 - 6	4 - 3	6 - 4	6 - 1
8 - 0	8 - 8	9 - 3	8 - 2
3 - 3	8 - 7	9 - 1	7 - 0
8 - 5	5 - 3	4 - 0	3 - 2
2 - 0	8 - 1	8 - 4	5 - 2

DAY 42
SUBTRACTING DIGITS
0–10

SCORE: | TIME: :

7 - 0	1 - 0	8 - 3	6 - 4
7 - 2	6 - 5	5 - 0	4 - 3
9 - 9	9 - 6	5 - 4	2 - 1
8 - 0	9 - 7	7 - 6	2 - 2
4 - 4	9 - 3	4 - 1	2 - 0
4 - 0	6 - 0	3 - 1	5 - 1
8 - 5	5 - 5	3 - 2	8 - 2
9 - 5	6 - 2	8 - 7	6 - 6
3 - 3	9 - 2	5 - 2	7 - 4
9 - 1	9 - 0	8 - 1	6 - 3

DAY 43

SUBTRACTING DIGITS 0–10

SCORE: | **TIME:** :

5 - 4	8 - 7	2 - 2	7 - 2
9 - 2	2 - 0	1 - 0	6 - 1
8 - 2	2 - 1	9 - 5	9 - 3
4 - 4	4 - 1	9 - 0	9 - 9
5 - 5	6 - 4	7 - 0	7 - 4
4 - 2	6 - 3	8 - 8	7 - 6
8 - 3	4 - 0	7 - 7	8 - 0
1 - 1	4 - 3	5 - 2	7 - 3
7 - 5	5 - 3	5 - 0	6 - 2
3 - 3	9 - 6	9 - 1	6 - 5

DAY 44

SUBTRACTING DIGITS

0–10

SCORE: | TIME: :

8 − 1 = ___	6 − 0 = ___	8 − 4 = ___	1 − 0 = ___
6 − 2 = ___	7 − 1 = ___	8 − 0 = ___	9 − 3 = ___
1 − 1 = ___	7 − 4 = ___	9 − 7 = ___	9 − 6 = ___
2 − 0 = ___	5 − 2 = ___	6 − 5 = ___	9 − 4 = ___
9 − 2 = ___	4 − 3 = ___	8 − 6 = ___	2 − 2 = ___
8 − 5 = ___	3 − 2 = ___	6 − 6 = ___	3 − 0 = ___
9 − 5 = ___	4 − 1 = ___	5 − 3 = ___	4 − 2 = ___
3 − 1 = ___	6 − 1 = ___	5 − 4 = ___	9 − 9 = ___
8 − 3 = ___	8 − 8 = ___	6 − 4 = ___	7 − 5 = ___
5 − 1 = ___	2 − 1 = ___	7 − 3 = ___	7 − 2 = ___

DAY 45

SUBTRACTING DIGITS

0–10

SCORE: | TIME: :

6 - 0	4 - 0	3 - 3	4 - 4
8 - 7	5 - 0	5 - 4	4 - 1
3 - 2	9 - 4	9 - 6	2 - 2
9 - 0	7 - 4	9 - 3	6 - 2
5 - 5	8 - 3	8 - 1	8 - 8
4 - 3	6 - 1	9 - 7	8 - 5
6 - 6	7 - 7	7 - 2	6 - 5
7 - 6	7 - 0	9 - 8	1 - 0
9 - 2	2 - 1	7 - 3	8 - 0
1 - 1	6 - 4	3 - 1	8 - 6

DAY 46

SUBTRACTING DIGITS

0–10

SCORE: | TIME: :

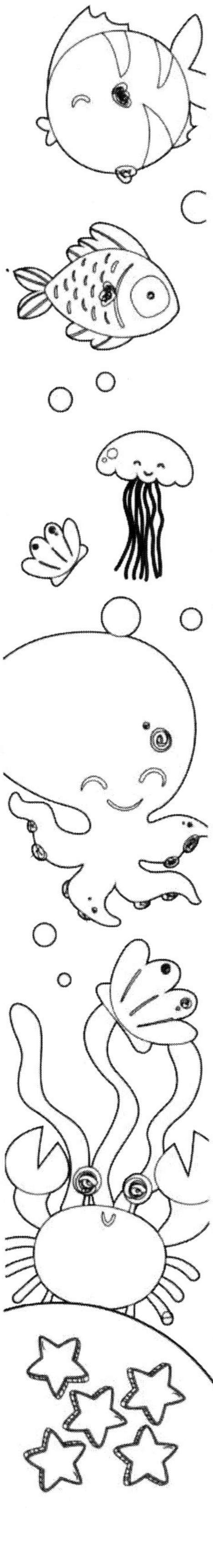

3 - 0	8 - 8	6 - 1	5 - 2
7 - 3	4 - 3	6 - 6	7 - 7
8 - 2	6 - 2	2 - 1	9 - 3
7 - 6	4 - 1	6 - 4	7 - 1
4 - 0	4 - 2	1 - 1	4 - 4
8 - 7	2 - 0	5 - 1	9 - 1
5 - 4	5 - 0	7 - 2	6 - 3
9 - 7	8 - 1	8 - 5	8 - 0
9 - 4	2 - 2	9 - 0	8 - 4
7 - 0	3 - 3	9 - 2	7 - 5

DAY 47

SUBTRACTING DIGITS 0–10

SCORE: | TIME: :

2 - 2	4 - 3	8 - 8	8 - 5
9 - 5	8 - 3	5 - 4	7 - 1
9 - 6	3 - 2	9 - 0	4 - 1
5 - 2	6 - 3	8 - 7	6 - 0
9 - 8	2 - 0	7 - 2	8 - 0
9 - 9	9 - 2	6 - 2	1 - 1
8 - 6	5 - 1	7 - 4	5 - 3
7 - 0	5 - 0	8 - 2	7 - 7
8 - 1	4 - 4	7 - 5	6 - 4
6 - 6	7 - 3	3 - 1	1 - 0

DAY 48
SUBTRACTING DIGITS 0–10

SCORE: | TIME: :

6 - 2	9 - 1	4 - 1	7 - 2
5 - 2	6 - 0	8 - 2	9 - 4
4 - 3	6 - 4	9 - 5	5 - 0
2 - 1	8 - 3	8 - 5	9 - 3
6 - 6	5 - 4	7 - 5	9 - 9
7 - 0	5 - 1	2 - 2	6 - 1
9 - 8	8 - 6	4 - 2	6 - 3
3 - 0	5 - 3	8 - 4	3 - 3
7 - 4	8 - 8	2 - 0	3 - 2
9 - 2	4 - 0	3 - 1	9 - 7

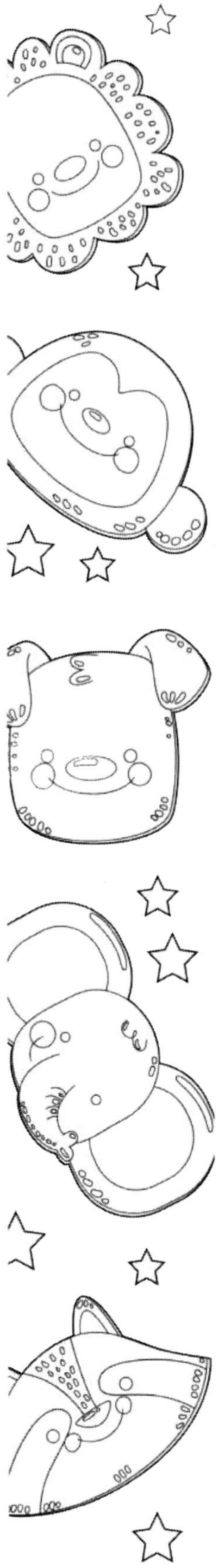

DAY 49

SUBTRACTING DIGITS

0–15

SCORE: | TIME: :

9 - 5	13 - 13	11 - 4	11 - 11
10 - 7	13 - 4	8 - 8	7 - 1
7 - 5	9 - 6	9 - 4	11 - 5
12 - 7	14 - 13	8 - 5	5 - 5
4 - 1	14 - 2	11 - 7	9 - 1
12 - 11	12 - 3	5 - 1	8 - 1
5 - 0	11 - 1	6 - 4	1 - 0
13 - 6	10 - 3	13 - 1	10 - 0
14 - 8	9 - 7	6 - 6	6 - 3
12 - 0	15 - 15	15 - 14	12 - 5

DAY 50

SUBTRACTING DIGITS

0–15

SCORE: TIME: :

12 - 12	14 - 1	14 - 10	13 - 1
11 - 6	2 - 2	10 - 0	5 - 4
14 - 8	13 - 3	3 - 0	14 - 9
9 - 5	13 - 12	7 - 6	11 - 7
10 - 4	14 - 3	10 - 3	8 - 0
10 - 7	11 - 5	4 - 1	12 - 10
6 - 1	3 - 2	6 - 4	13 - 10
15 - 2	15 - 13	9 - 8	12 - 4
10 - 1	10 - 2	13 - 0	14 - 4
14 - 7	14 - 0	12 - 3	7 - 1

DAY 51

SUBTRACTING DIGITS

0–15

SCORE: | **TIME:** :

15 - 12	12 - 0	5 - 4	15 - 0
11 - 8	5 - 5	15 - 5	14 - 10
15 - 3	6 - 5	10 - 10	11 - 4
14 - 0	2 - 1	8 - 2	8 - 5
9 - 7	3 - 3	3 - 2	13 - 5
12 - 3	14 - 1	12 - 6	2 - 0
9 - 1	4 - 2	3 - 0	10 - 6
9 - 9	5 - 2	10 - 3	6 - 6
11 - 0	9 - 6	15 - 11	7 - 1
3 - 1	15 - 9	10 - 4	8 - 1

DAY 52

SUBTRACTING DIGITS

0–15

SCORE: | TIME: :

15 - 3	7 - 7	11 - 1	14 - 8
15 - 0	13 - 3	7 - 3	12 - 7
8 - 4	12 - 5	14 - 10	10 - 7
14 - 13	15 - 1	2 - 2	14 - 12
14 - 3	11 - 9	14 - 4	13 - 4
15 - 8	12 - 4	12 - 10	13 - 2
12 - 2	15 - 5	11 - 11	11 - 5
11 - 4	10 - 8	11 - 0	10 - 6
13 - 9	14 - 6	14 - 14	13 - 12
1 - 0	13 - 11	5 - 1	15 - 6

DAY 53

SUBTRACTING DIGITS

0–15

SCORE: | TIME: :

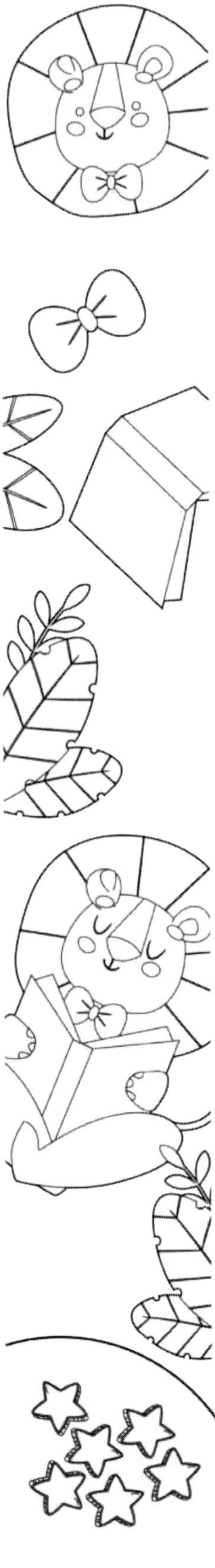

13 - 3	10 - 8	5 - 2	15 - 4
10 - 5	14 - 10	12 - 1	15 - 10
6 - 3	10 - 9	8 - 5	13 - 4
2 - 0	6 - 1	10 - 3	10 - 4
15 - 2	9 - 2	12 - 9	10 - 1
9 - 7	14 - 13	12 - 12	4 - 1
12 - 7	13 - 13	6 - 2	8 - 2
11 - 11	11 - 1	14 - 9	8 - 1
13 - 0	7 - 7	7 - 1	12 - 0
4 - 4	10 - 10	4 - 2	6 - 5

DAY 54

SUBTRACTING DIGITS

0–15

SCORE: | TIME: :

11 - 4	10 - 7	8 - 4	11 - 2
13 - 0	6 - 0	12 - 1	7 - 1
6 - 4	14 - 14	12 - 11	12 - 4
11 - 3	14 - 5	11 - 1	3 - 3
13 - 9	15 - 15	15 - 8	15 - 4
15 - 10	12 - 7	11 - 11	13 - 10
8 - 1	1 - 1	9 - 7	14 - 3
7 - 2	2 - 2	12 - 8	10 - 8
10 - 5	15 - 1	10 - 4	4 - 3
15 - 11	14 - 11	7 - 3	9 - 1

DAY 55

SUBTRACTING DIGITS

0–15

SCORE: | TIME: :

2 − 1	10 − 0	14 − 4	14 − 10
12 − 2	12 − 9	14 − 0	11 − 3
13 − 1	13 − 9	13 − 0	7 − 3
13 − 12	3 − 0	15 − 12	6 − 6
15 − 9	5 − 1	15 − 11	1 − 0
14 − 14	12 − 0	3 − 2	13 − 2
13 − 6	14 − 6	11 − 11	9 − 0
9 − 4	12 − 7	10 − 9	4 − 4
10 − 2	8 − 8	2 − 0	7 − 1
13 − 4	12 − 6	14 − 13	5 − 3

DAY 56

SUBTRACTING DIGITS 0–20

SCORE: | TIME: :

5 - 0	16 - 15	16 - 0	10 - 5
7 - 2	17 - 4	3 - 3	20 - 12
17 - 0	10 - 10	9 - 5	5 - 4
18 - 8	8 - 7	16 - 2	4 - 4
7 - 7	18 - 5	20 - 20	16 - 11
18 - 13	12 - 4	4 - 3	15 - 9
11 - 10	17 - 1	20 - 15	14 - 5
19 - 1	13 - 11	15 - 10	19 - 5
15 - 0	15 - 5	14 - 0	13 - 7
19 - 7	7 - 4	10 - 1	2 - 2

DAY 57

SUBTRACTING DIGITS 0–20

SCORE: | TIME: :

11 − 1	16 − 14	4 − 4	19 − 11
14 − 11	12 − 11	17 − 6	12 − 5
18 − 14	2 − 2	20 − 16	18 − 17
7 − 7	16 − 11	14 − 1	20 − 18
16 − 8	12 − 12	17 − 14	15 − 11
9 − 0	17 − 17	12 − 7	16 − 10
16 − 0	17 − 10	19 − 14	15 − 1
18 − 3	12 − 3	3 − 3	2 − 0
19 − 18	8 − 6	11 − 0	7 − 4
13 − 12	10 − 2	14 − 12	20 − 4

DAY 58

SUBTRACTING DIGITS 0–20

SCORE: | TIME: :

18 - 7	20 - 17	19 - 9	8 - 0
20 - 3	16 - 4	5 - 2	20 - 18
19 - 0	20 - 6	15 - 9	12 - 2
11 - 0	8 - 2	15 - 5	19 - 15
16 - 15	17 - 9	14 - 3	13 - 4
17 - 17	13 - 10	3 - 2	18 - 16
14 - 13	10 - 0	19 - 19	7 - 1
20 - 10	17 - 5	18 - 9	14 - 14
20 - 7	6 - 6	16 - 12	15 - 4
16 - 9	19 - 14	13 - 9	11 - 1

DAY 59

SUBTRACTING DIGITS

0–20

SCORE: | TIME: :

15 - 11	6 - 5	8 - 3	17 - 3
5 - 2	10 - 8	19 - 19	17 - 1
20 - 8	20 - 6	14 - 2	10 - 0
15 - 13	2 - 1	10 - 1	8 - 6
13 - 2	20 - 3	4 - 4	20 - 5
20 - 9	7 - 4	17 - 14	14 - 7
18 - 10	20 - 10	14 - 13	19 - 4
13 - 13	19 - 6	13 - 0	13 - 9
13 - 3	15 - 8	8 - 8	18 - 4
16 - 12	11 - 10	5 - 4	18 - 15

DAY 60

SUBTRACTING DIGITS 0–20

SCORE: | TIME: :

16 - 16	15 - 4	12 - 2	13 - 8
7 - 0	20 - 2	17 - 0	16 - 7
16 - 9	8 - 6	16 - 1	3 - 3
18 - 1	7 - 6	15 - 5	20 - 9
10 - 2	11 - 4	14 - 8	16 - 8
20 - 6	16 - 0	20 - 12	3 - 2
16 - 3	10 - 4	8 - 4	17 - 3
5 - 0	16 - 15	14 - 14	6 - 1
13 - 9	7 - 2	14 - 5	16 - 2
17 - 8	20 - 4	19 - 16	9 - 5

DAY 61

SUBTRACTING DIGITS
0–20

SCORE: | TIME: :

10 - 0	16 - 2	10 - 7	20 - 16
5 - 0	16 - 0	16 - 6	20 - 5
15 - 1	17 - 11	4 - 2	20 - 17
13 - 1	15 - 11	14 - 1	9 - 6
18 - 9	17 - 16	10 - 3	15 - 12
11 - 8	17 - 0	17 - 4	5 - 3
19 - 8	13 - 0	6 - 6	9 - 8
12 - 4	4 - 3	10 - 1	15 - 14
17 - 12	14 - 3	6 - 2	9 - 1
16 - 16	12 - 2	8 - 2	17 - 5

DAY 62

SUBTRACTING DIGITS 0–20

SCORE: | TIME: :

8 − 0	16 − 1	14 − 12	17 − 4
14 − 5	13 − 1	20 − 6	19 − 9
17 − 11	20 − 0	19 − 8	9 − 0
2 − 1	18 − 6	19 − 11	20 − 12
9 − 8	7 − 3	7 − 0	16 − 4
11 − 0	16 − 11	10 − 3	17 − 1
8 − 7	14 − 9	12 − 1	20 − 20
11 − 10	18 − 9	16 − 16	8 − 3
11 − 1	14 − 4	10 − 7	16 − 2
17 − 2	20 − 5	18 − 5	20 − 4

DAY 63
SUBTRACTING DIGITS
0–20

SCORE: | TIME: :

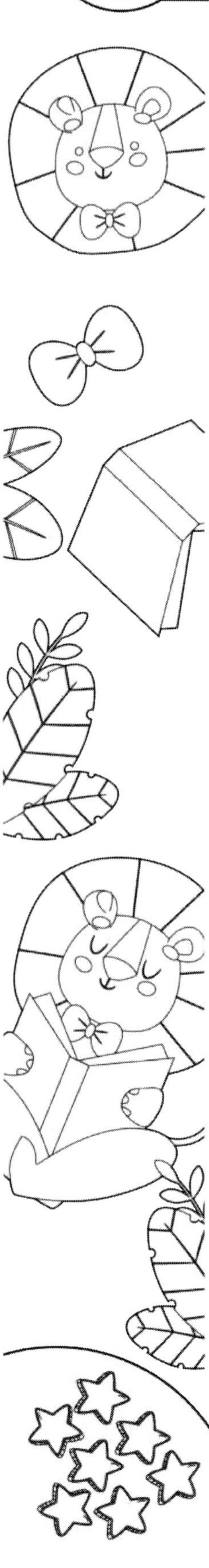

16 - 14	12 - 10	15 - 9	2 - 2
6 - 5	10 - 0	17 - 14	11 - 1
7 - 5	16 - 4	8 - 8	14 - 1
17 - 8	7 - 6	8 - 6	6 - 1
5 - 4	5 - 3	19 - 2	16 - 11
13 - 2	18 - 11	10 - 3	18 - 4
4 - 4	17 - 3	15 - 14	16 - 12
15 - 1	19 - 13	20 - 19	6 - 0
3 - 3	18 - 10	16 - 3	18 - 16
16 - 5	5 - 5	11 - 9	19 - 9

DAY 64
SUBTRACTING DIGITS
0–20

SCORE: | TIME: :

17 - 15	6 - 0	18 - 17	20 - 4
9 - 6	20 - 12	9 - 9	14 - 5
18 - 2	19 - 9	8 - 5	14 - 0
7 - 5	18 - 18	19 - 7	16 - 11
19 - 17	17 - 7	19 - 5	20 - 20
7 - 4	13 - 7	11 - 2	17 - 0
19 - 16	13 - 2	10 - 9	14 - 12
14 - 1	20 - 10	9 - 4	18 - 4
17 - 9	16 - 8	8 - 4	13 - 13
7 - 2	19 - 15	15 - 7	15 - 15

DAY 65

SUBTRACTING DIGITS

0–20

SCORE: | TIME: :

17 - 6	6 - 1	19 - 10	19 - 0
20 - 12	17 - 5	10 - 0	15 - 6
19 - 8	9 - 4	18 - 5	8 - 2
11 - 3	17 - 2	14 - 6	15 - 0
10 - 9	18 - 12	5 - 5	9 - 9
12 - 9	2 - 2	20 - 0	15 - 10
7 - 5	19 - 3	19 - 9	9 - 0
17 - 3	10 - 6	4 - 3	2 - 0
3 - 1	17 - 14	6 - 4	12 - 10
20 - 15	19 - 6	15 - 14	13 - 5

DAY 66
SUBTRACTING DIGITS
0–20

SCORE: | TIME: :

15 - 3	13 - 1	3 - 3	11 - 3
13 - 3	18 - 10	14 - 11	18 - 16
7 - 4	15 - 7	19 - 16	14 - 13
2 - 2	17 - 0	7 - 2	20 - 14
13 - 7	8 - 1	17 - 17	18 - 9
13 - 11	15 - 6	8 - 6	15 - 14
7 - 5	12 - 8	8 - 7	13 - 10
19 - 9	6 - 2	10 - 3	19 - 1
14 - 4	5 - 3	18 - 17	18 - 7
10 - 2	19 - 17	13 - 12	15 - 1

DAY 67

SUBTRACTING DIGITS 0–20

SCORE: | TIME: :

20 - 5	5 - 5	9 - 6	17 - 0
17 - 15	10 - 0	10 - 2	12 - 2
10 - 7	18 - 16	19 - 13	20 - 16
18 - 1	12 - 6	7 - 4	16 - 15
20 - 2	14 - 7	15 - 13	15 - 3
18 - 5	12 - 9	12 - 12	17 - 9
9 - 7	1 - 1	4 - 4	20 - 7
18 - 4	3 - 3	13 - 6	20 - 12
18 - 6	17 - 6	20 - 1	15 - 15
8 - 5	14 - 10	18 - 12	10 - 8

DAY 68

SUBTRACTING DIGITS

0–20

SCORE: | TIME: :

10 − 6 =	5 − 0 =	16 − 11 =	14 − 13 =
9 − 0 =	16 − 0 =	20 − 11 =	4 − 1 =
7 − 2 =	15 − 14 =	7 − 5 =	12 − 12 =
20 − 4 =	16 − 1 =	19 − 18 =	8 − 8 =
2 − 0 =	6 − 6 =	10 − 2 =	19 − 0 =
13 − 8 =	13 − 4 =	19 − 4 =	19 − 2 =
17 − 8 =	18 − 5 =	12 − 11 =	19 − 13 =
14 − 12 =	20 − 1 =	17 − 1 =	13 − 2 =
13 − 13 =	16 − 16 =	7 − 7 =	20 − 20 =
11 − 10 =	19 − 5 =	12 − 5 =	11 − 1 =

DAY 69

SUBTRACTING DIGITS

0–20

SCORE: | TIME: :

19 - 14	11 - 10	7 - 3	14 - 11
18 - 16	17 - 4	9 - 7	19 - 6
14 - 14	14 - 3	14 - 2	13 - 2
18 - 5	10 - 0	8 - 1	17 - 16
13 - 10	14 - 9	13 - 6	13 - 4
20 - 10	18 - 11	17 - 1	12 - 12
19 - 7	20 - 17	20 - 7	15 - 9
16 - 8	17 - 10	17 - 13	17 - 12
20 - 5	13 - 5	14 - 10	10 - 5
12 - 8	20 - 2	14 - 13	15 - 5

DAY 70

SUBTRACTING DIGITS

0–20

SCORE: | TIME: :

16 - 9	11 - 4	16 - 4	19 - 15
14 - 8	13 - 2	20 - 4	10 - 2
19 - 16	10 - 5	19 - 19	16 - 2
13 - 6	20 - 5	17 - 14	4 - 1
2 - 2	16 - 12	18 - 0	2 - 0
7 - 6	6 - 4	8 - 8	16 - 15
19 - 1	9 - 9	6 - 3	4 - 2
17 - 7	10 - 9	14 - 11	17 - 5
16 - 8	18 - 10	10 - 0	16 - 7
17 - 16	15 - 8	11 - 9	13 - 12

DAY 71

SUBTRACTING DIGITS 0–20

SCORE: | TIME: :

17 - 8	20 - 6	16 - 9	10 - 7
9 - 4	14 - 7	12 - 1	13 - 8
20 - 7	17 - 10	9 - 7	18 - 9
20 - 8	16 - 10	13 - 7	6 - 2
15 - 8	14 - 13	19 - 12	16 - 11
20 - 8	18 - 0	19 - 1	9 - 1
20 - 9	13 - 0	8 - 8	7 - 6
13 - 4	20 - 20	10 - 7	19 - 4
15 - 14	16 - 11	8 - 8	4 - 2
13 - 12	15 - 7	3 - 2	18 - 4

DAY 72

SUBTRACTING DIGITS

0–20

SCORE: | TIME: :

12 − 9	7 − 6	14 − 9	16 − 0
20 − 16	16 − 6	4 − 3	5 − 5
8 − 1	14 − 14	3 − 2	18 − 14
14 − 7	9 − 2	7 − 5	10 − 10
3 − 3	20 − 5	17 − 0	16 − 9
20 − 20	12 − 1	12 − 12	14 − 2
19 − 19	11 − 10	4 − 4	7 − 2
7 − 4	20 − 1	18 − 16	9 − 7
8 − 7	10 − 1	13 − 7	6 − 0
12 − 7	10 − 8	19 − 1	9 − 6

DAY

SUBTRACTING DIGITS

0–20

73

SCORE: | **TIME:** :

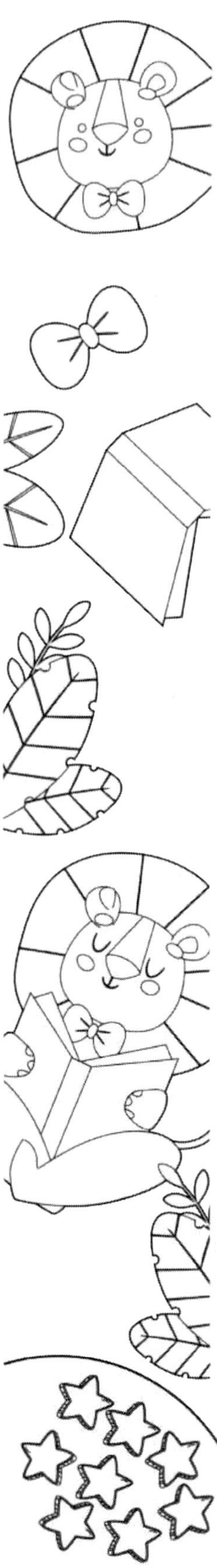

16 - 4	8 - 0	19 - 14	15 - 5
11 - 9	18 - 8	8 - 7	14 - 1
13 - 11	4 - 1	5 - 1	15 - 13
16 - 13	13 - 4	18 - 9	10 - 2
6 - 5	12 - 12	19 - 5	20 - 17
4 - 4	20 - 3	14 - 8	8 - 6
4 - 0	20 - 11	20 - 1	20 - 2
18 - 18	17 - 13	12 - 0	9 - 9
14 - 0	19 - 10	9 - 7	10 - 4
9 - 5	19 - 18	12 - 10	17 - 3

DAY 74
SUBTRACTING DIGITS
0–20

SCORE: | TIME: :

16 - 11	17 - 4	20 - 6	20 - 18
12 - 12	11 - 9	16 - 9	9 - 7
1 - 0	19 - 17	10 - 8	9 - 1
19 - 14	20 - 11	16 - 5	19 - 6
18 - 15	20 - 7	17 - 2	9 - 6
11 - 1	16 - 13	18 - 17	20 - 8
14 - 4	6 - 2	20 - 14	17 - 8
10 - 4	7 - 7	14 - 9	13 - 6
19 - 3	11 - 8	9 - 4	10 - 7
7 - 4	5 - 1	7 - 1	15 - 12

DAY 75
SUBTRACTING DIGITS
0–20

SCORE: | TIME: :

10 − 3	17 − 4	19 − 13	19 − 12
6 − 5	19 − 1	13 − 2	19 − 7
18 − 16	10 − 2	11 − 8	15 − 1
6 − 3	7 − 0	9 − 9	20 − 17
10 − 7	9 − 6	12 − 4	7 − 3
9 − 4	4 − 4	15 − 5	11 − 6
13 − 0	17 − 0	9 − 2	20 − 18
8 − 3	11 − 7	16 − 8	4 − 2
16 − 14	6 − 6	11 − 10	12 − 1
17 − 10	19 − 8	6 − 0	12 − 3

DAY 76

SUBTRACTING DIGITS

0–20

SCORE: | TIME: :

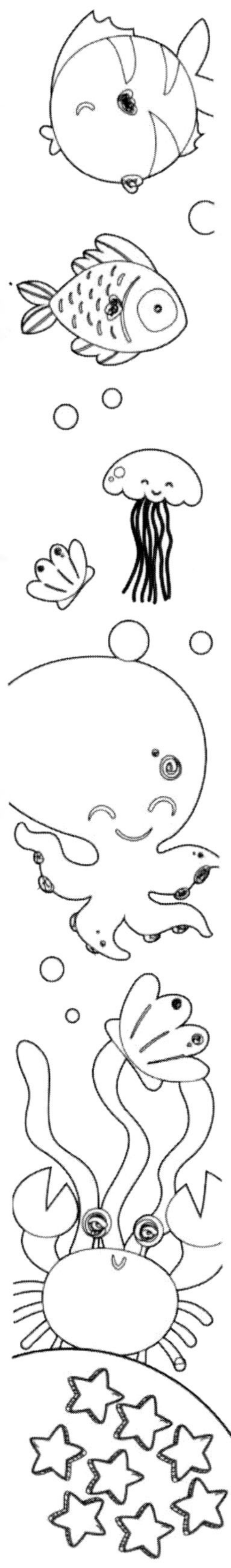

18 − 1	7 − 0	20 − 15	13 − 0
8 − 5	14 − 13	11 − 11	10 − 10
20 − 8	17 − 13	12 − 9	7 − 6
8 − 0	3 − 3	7 − 2	2 − 2
3 − 2	9 − 8	12 − 6	20 − 10
11 − 5	14 − 8	15 − 5	8 − 6
9 − 6	11 − 9	6 − 0	13 − 5
18 − 3	17 − 7	15 − 11	13 − 9
19 − 18	7 − 3	9 − 1	14 − 0
16 − 2	10 − 0	17 − 8	19 − 16

DAY 77

SUBTRACTING DIGITS

0–20

SCORE: | TIME: :

18 - 13	18 - 12	13 - 12	8 - 4
16 - 9	16 - 15	7 - 4	16 - 16
13 - 3	7 - 5	12 - 1	18 - 10
8 - 7	20 - 14	6 - 6	19 - 15
14 - 11	13 - 11	4 - 4	14 - 0
17 - 13	8 - 3	10 - 3	20 - 5
11 - 2	3 - 2	16 - 13	15 - 3
20 - 3	14 - 7	15 - 12	16 - 6
14 - 5	5 - 5	12 - 10	20 - 2
13 - 2	20 - 0	13 - 13	8 - 0

DAY 78

SUBTRACTING DIGITS 0–20

SCORE: | TIME: :

18 - 16	15 - 1	19 - 6	5 - 2
7 - 6	15 - 0	19 - 1	9 - 9
18 - 6	10 - 3	16 - 4	19 - 12
10 - 9	14 - 10	1 - 0	18 - 14
17 - 15	11 - 3	17 - 3	15 - 2
8 - 0	9 - 4	19 - 10	13 - 11
19 - 9	12 - 8	12 - 0	15 - 10
20 - 9	11 - 4	17 - 16	17 - 6
7 - 5	19 - 3	19 - 15	17 - 8
20 - 14	7 - 4	4 - 1	16 - 16

DAY 79

SUBTRACTING DIGITS 0–20

SCORE: | TIME: :

11 - 2	8 - 0	11 - 3	19 - 15
18 - 11	18 - 3	16 - 6	19 - 0
20 - 11	20 - 9	19 - 7	12 - 10
12 - 8	18 - 9	10 - 8	19 - 14
15 - 12	19 - 1	6 - 5	9 - 5
8 - 3	20 - 17	14 - 2	5 - 1
18 - 12	7 - 4	10 - 9	4 - 3
9 - 6	4 - 1	20 - 1	19 - 6
8 - 8	7 - 7	16 - 8	16 - 0
16 - 9	12 - 1	17 - 4	19 - 2

DAY 80

SUBTRACTING DIGITS

0–20

SCORE: | TIME: :

20 − 20	6 − 2	13 − 2	5 − 4
14 − 2	18 − 6	15 − 13	17 − 11
8 − 8	15 − 9	18 − 14	13 − 8
11 − 0	10 − 2	18 − 5	7 − 5
4 − 2	13 − 5	19 − 10	11 − 3
9 − 8	9 − 6	20 − 18	15 − 12
7 − 2	5 − 0	13 − 7	17 − 14
11 − 5	12 − 7	11 − 8	17 − 12
9 − 7	15 − 2	1 − 1	12 − 12
10 − 8	19 − 8	14 − 0	19 − 7

DAY 81

ADDING ANDSUBTRACTING

0–20

SCORE: | TIME: :

9 + 7	9 - 0	10 - 9	13 + 6	10 + 9
1 + 1	1 + 10	15 + 12	5 + 15	7 - 5
7 + 0	5 + 14	7 - 4	16 + 12	13 - 11
11 - 2	16 + 17	12 - 0	5 + 11	16 - 12
4 + 1	14 - 13	15 - 15	19 - 2	19 - 10
5 + 9	11 + 6	10 + 10	3 - 2	14 - 7
14 + 9	19 + 15	19 - 5	15 - 7	17 - 11
11 - 8	17 - 4	16 - 1	7 + 4	10 - 5

DAY 82

ADDING ANDSUBTRACTING

0–20

SCORE: | TIME: :

14 + 7	4 + 11	14 - 13	9 + 8	11 - 1
3 + 6	6 + 4	4 + 6	15 + 3	9 + 1
4 - 0	18 + 11	3 + 5	11 - 9	5 - 5
10 + 16	0 + 7	15 + 2	9 - 6	3 + 6
15 - 1	13 + 16	9 - 3	0 + 16	6 - 4
13 + 7	2 + 7	10 + 12	1 + 16	6 - 5
3 - 2	2 - 0	10 - 6	13 + 16	11 - 11
16 - 4	16 + 18	18 - 5	9 - 2	17 + 17

DAY 83

ADDING ANDSUBTRACTING

0–20

SCORE: | TIME: :

12 - 9	16 - 4	9 + 15	18 + 3	18 - 12
5 + 1	18 - 0	11 - 9	4 + 5	3 + 11
14 - 14	9 + 13	17 - 17	1 + 5	17 - 12
17 + 14	14 - 6	8 + 10	12 - 2	17 + 11
15 + 17	12 - 6	11 - 2	5 + 4	4 + 7
12 + 14	17 + 12	18 - 14	19 + 2	17 - 11
16 + 14	14 - 11	17 - 5	19 - 6	12 - 8
18 - 6	14 + 5	16 - 15	12 - 7	17 - 15

DAY 84

ADDING AND SUBTRACTING

0–20

SCORE: | TIME: :

16 + 2	10 + 15	19 - 6	3 + 15	15 - 7
7 + 10	13 - 13	1 + 16	8 + 4	18 + 5
16 - 3	17 - 7	18 - 0	18 + 17	18 - 15
14 + 17	13 + 18	2 + 4	8 - 1	4 + 13
9 - 6	0 + 13	5 + 0	7 + 4	11 - 3
18 - 4	9 - 7	8 + 13	13 - 6	2 + 9
1 + 13	3 - 2	14 + 6	7 + 4	18 - 4
14 + 8	15 + 2	16 + 12	10 - 8	11 - 6

DAY 85

ADDING ANDSUBTRACTING

0–20

SCORE: | TIME: :

9 - 9	15 - 1	9 + 12	6 + 12	13 - 3
15 - 3	16 + 16	16 - 16	12 - 4	6 + 8
18 + 12	8 - 2	6 + 6	8 - 6	16 - 7
3 + 7	12 - 4	7 + 14	18 - 15	4 - 3
17 + 18	9 - 4	11 + 18	10 + 12	10 - 9
3 + 3	17 - 6	5 + 13	15 + 8	13 + 6
9 - 5	7 - 3	14 + 13	11 + 10	13 - 3
5 - 2	1 + 18	14 + 11	10 + 15	9 + 0

DAY 86

ADDING ANDSUBTRACTING

0–20

SCORE: | TIME: :

16 + 12	18 + 1	16 + 11	3 + 4	18 + 13
2 + 12	19 - 11	9 - 3	9 + 4	17 - 5
3 + 16	7 - 1	9 - 1	11 - 5	17 - 11
15 + 8	3 + 2	14 + 18	13 + 17	13 - 2
18 - 18	9 + 10	1 + 0	19 - 8	8 - 4
12 - 4	6 - 5	5 - 3	18 + 7	7 - 3
7 - 2	17 - 3	17 - 8	12 - 11	5 + 10
0 + 2	15 - 7	18 - 8	15 + 0	18 - 17

DAY 87

ADDING ANDSUBTRACTING

0–20

SCORE: | TIME: :

14	15	10	6	11
- 8	+ 15	+ 9	- 2	- 8
10	18	12	8	10
- 6	- 17	- 10	- 3	+ 6
16	6	11	10	3
- 7	+ 7	- 10	+ 13	+ 4
13	12	15	16	8
+ 5	- 7	- 12	+ 18	- 3
12	17	15	1	19
- 9	- 5	+ 11	+ 4	- 4
14	16	11	6	3
+ 16	+ 3	- 7	- 1	+ 6
16	9	17	16	7
+ 10	+ 11	- 14	- 8	- 4
14	14	9	17	16
- 11	- 6	+ 10	+ 15	+ 5

DAY 88

ADDING ANDSUBTRACTING

0–20

SCORE: | TIME: :

20 + 1	10 + 13	10 - 4	20 - 1	10 - 6
10 + 1	20 - 6	10 + 18	20 - 7	10 + 12
10 - 0	10 - 6	10 + 10	20 + 1	10 - 3
10 - 3	10 + 4	10 - 8	10 + 8	10 - 3
10 - 4	20 - 12	10 + 2	20 - 2	10 - 3
20 - 13	10 + 2	10 - 13	10 - 4	20 + 3
10 - 1	20 + 17	10 - 1	10 - 2	10 - 8
20 - 5	10 + 15	10 + 16	10 - 2	10 - 12

DAY 89

ADDING ANDSUBTRACTING

0–20

SCORE: | TIME: :

10 + 15	20 - 15	10 - 1	10 - 2	10 + 4
10 - 3	10 + 10	10 - 9	10 + 10	20 + 1
10 + 6	10 - 10	10 + 1	20 + 3	20 + 5
10 + 19	10 + 3	20 - 7	10 + 10	20 + 4
10 - 3	10 + 12	10 - 1	10 + 7	10 + 9
10 - 4	10 + 19	10 + 1	10 + 4	10 - 5
10 + 15	20 + 11	10 + 8	20 - 14	10 - 6
20 - 4	10 - 7	20 - 13	20 - 11	10 - 2

DAY 90

ADDING ANDSUBTRACTING

0–20

SCORE: | TIME: :

20 + 3	20 - 7	10 + 4	20 - 16	10 - 2
10 + 7	20 + 16	10 - 4	10 + 15	20 - 1
10 + 13	10 + 19	10 + 11	10 + 5	10 + 10
10 - 9	10 + 8	20 - 1	10 + 2	10 - 10
20 - 6	10 + 8	10 + 10	10 - 8	20 - 8
10 - 0	20 - 1	10 - 1	20 + 14	10 - 0
10 - 4	20 - 2	10 + 15	10 - 2	20 - 7
10 - 5	20 - 11	10 + 5	20 - 15	20 + 8

DAY 91

ADDING ANDSUBTRACTING

0–20

SCORE: | TIME: :

10 - 5	10 - 4	10 - 4	20 - 2	20 + 18
10 + 12	20 + 17	10 + 5	10 - 2	10 + 1
20 - 1	10 + 9	10 - 8	10 + 11	10 + 8
10 - 3	10 - 11	10 - 5	10 - 1	10 + 10
10 - 12	10 + 9	20 - 12	20 + 8	10 + 7
10 + 16	10 - 10	20 - 10	10 - 8	20 - 9
10 + 1	10 + 8	10 + 13	20 - 9	10 - 1
20 - 9	20 - 9	10 - 3	10 + 10	20 - 6

DAY 92

ADDING ANDSUBTRACTING

0–20

SCORE: | TIME: :

13 + 1	1 + 10	18 - 6	19 - 3	12 - 9
15 - 13	13 + 12	6 - 5	12 - 1	4 + 4
2 - 1	15 + 6	19 - 7	2 + 12	4 + 0
11 + 15	17 - 7	17 - 8	14 - 14	17 + 7
18 - 4	6 - 5	2 + 3	17 - 15	4 + 7
5 - 0	18 + 10	8 - 4	4 + 7	10 + 13
7 + 17	19 - 5	1 + 4	6 + 9	19 - 9
11 + 1	9 - 6	5 + 15	16 - 8	12 + 7

DAY 93

ADDING ANDSUBTRACTING

0–20

SCORE: | TIME: :

3 + 3	5 + 7	15 + 7	16 + 2	16 - 1
2 - 2	18 + 12	7 + 16	12 - 3	12 - 6
7 - 3	5 - 5	4 - 2	8 - 5	17 + 7
1 + 11	0 + 10	14 - 9	16 - 12	3 + 16
4 + 17	8 - 6	14 - 1	6 + 14	16 - 2
16 - 11	17 - 2	18 - 0	12 - 1	10 + 18
5 - 2	19 - 17	1 + 5	14 - 1	8 - 6
8 - 0	7 + 1	7 + 2	17 - 7	7 + 4

DAY 94

ADDING ANDSUBTRACTING

0–20

SCORE: | TIME: :

20 - 12	20 - 14	10 + 2	10 - 1	10 - 6
10 - 2	10 + 0	10 + 18	20 - 4	10 - 7
20 + 11	20 - 7	20 - 9	10 - 12	20 - 2
10 + 13	10 + 4	20 + 2	10 - 10	10 + 5
10 - 1	10 - 3	20 - 2	20 - 4	20 - 10
10 + 5	10 - 4	20 - 4	20 - 13	20 - 12
20 - 4	10 - 6	10 + 18	10 + 8	10 + 5
20 - 3	10 + 13	10 - 3	20 - 4	10 - 6

DAY 95

ADDING ANDSUBTRACTING

0–20

SCORE: | TIME: :

17 + 14	17 - 10	12 + 8	15 + 11
8 - 3	15 - 7	17 - 2	2 + 3
8 - 7	2 + 6	3 + 1	5 + 18
11 - 1	18 - 1	7 + 11	19 - 14
10 - 6	9 + 5	19 - 9	6 - 1
16 - 9	12 - 8	1 + 5	7 + 5

DAY 96

ADDING ANDSUBTRACTING

0–20

SCORE: | TIME: :

13 - 8	12 - 7	18 - 3	7 + 5
6 + 16	8 - 8	11 + 9	1 + 12
14 - 8	17 - 12	6 - 1	16 - 7
13 - 7	4 + 18	19 + 7	14 - 0
11 + 1	10 - 6	10 - 0	6 + 9
11 + 1	3 + 17	17 - 6	19 - 12

DAY 97

ADDING ANDSUBTRACTING

0–20

SCORE: | TIME: :

10 - 5	3 + 18	17 - 4	7 - 1
13 - 11	18 - 12	8 + 0	18 - 14
6 - 2	12 - 2	16 + 4	7 - 4
18 - 2	15 - 5	10 - 9	17 + 5
18 + 7	0 - 0	8 + 5	8 + 16
15 + 9	10 - 0	15 + 7	15 - 5

DAY 98

ADDING ANDSUBTRACTING

0–20

SCORE: | TIME: :

12 - 3	15 - 9	7 + 12	10 + 7
7 + 7	4 + 1	13 - 10	12 - 2
8 + 2	11 + 5	6 + 5	11 - 2
13 + 9	1 + 17	16 - 13	16 - 7
16 - 6	9 + 8	16 - 14	13 - 5
5 + 7	18 - 1	6 + 2	16 + 9

DAY 99

ADDING ANDSUBTRACTING

0–20

SCORE: | TIME: :

6 + 9	7 + 0	11 - 1	9 + 16
10 - 1	1 + 7	12 + 1	17 - 12
7 - 1	17 - 1	11 + 1	5 + 11
15 - 10	4 + 5	16 + 18	18 + 8
11 - 7	16 + 2	13 + 12	1 + 14
17 + 3	15 - 5	18 - 6	12 - 12

DAY 100

ADDING ANDSUBTRACTING

0–20

SCORE: | TIME: :

14 - 14	14 + 15	18 - 3	10 - 8
6 + 17	12 + 13	11 + 9	7 - 1
9 - 4	16 - 5	8 - 4	2 - 1
9 - 0	14 + 3	15 - 2	10 - 8
18 - 8	17 - 9	19 - 6	6 - 6
6 + 10	3 + 7	15 + 16	10 + 3

ANSWER DAY 1-4

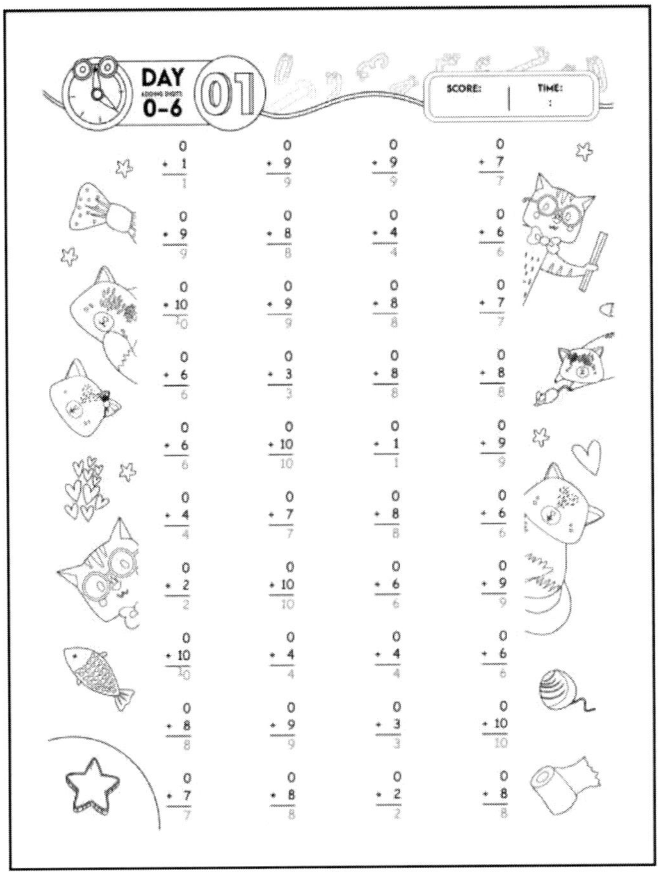

DAY 01 ADDING DIGITS 0-6 SCORE: TIME:

0 + 1 = 1	0 + 9 = 9	0 + 9 = 9	0 + 7 = 7
0 + 9 = 9	0 + 8 = 8	0 + 4 = 4	0 + 6 = 6
0 + 10 = 10	0 + 9 = 9	0 + 8 = 8	0 + 7 = 7
0 + 6 = 6	0 + 3 = 3	0 + 8 = 8	0 + 8 = 8
0 + 6 = 6	0 + 10 = 10	0 + 1 = 1	0 + 9 = 9
0 + 4 = 4	0 + 7 = 7	0 + 8 = 8	0 + 6 = 6
0 + 2 = 2	0 + 10 = 10	0 + 6 = 6	0 + 9 = 9
0 + 10 = 10	0 + 4 = 4	0 + 4 = 4	0 + 6 = 6
0 + 8 = 8	0 + 9 = 9	0 + 3 = 3	0 + 10 = 10
0 + 7 = 7	0 + 8 = 8	0 + 2 = 2	0 + 8 = 8

DAY 02 ADDING DIGITS 0-6 SCORE: TIME:

1 + 0 = 1	1 + 8 = 9	1 + 9 = 10	1 + 7 = 8
1 + 1 = 2	1 + 6 = 7	1 + 3 = 4	1 + 8 = 9
1 + 9 = 10	1 + 7 = 8	1 + 6 = 7	1 + 4 = 5
1 + 8 = 9	1 + 10 = 11	1 + 5 = 6	1 + 3 = 4
1 + 9 = 10	1 + 9 = 10	1 + 9 = 10	1 + 4 = 5
1 + 10 = 11	1 + 0 = 1	1 + 10 = 11	1 + 0 = 1
1 + 10 = 11	1 + 4 = 5	1 + 7 = 8	1 + 5 = 6
1 + 5 = 6	1 + 8 = 9	1 + 9 = 10	1 + 7 = 8
1 + 3 = 4	1 + 4 = 5	1 + 7 = 8	1 + 9 = 10
1 + 2 = 3	1 + 5 = 6	1 + 0 = 1	1 + 2 = 3

DAY 03 ADDING DIGITS 0-6 SCORE: TIME:

2 + 7 = 9	2 + 0 = 2	2 + 4 = 6	2 + 5 = 7
2 + 1 = 3	2 + 3 = 5	2 + 10 = 12	2 + 3 = 5
2 + 4 = 6	2 + 0 = 2	2 + 7 = 9	2 + 0 = 2
2 + 10 = 12	2 + 3 = 5	2 + 7 = 9	2 + 1 = 3
2 + 0 = 2	2 + 3 = 5	2 + 2 = 4	2 + 3 = 5
2 + 0 = 2	2 + 1 = 3	2 + 3 = 5	2 + 4 = 6
2 + 8 = 10	2 + 2 = 4	2 + 2 = 4	2 + 1 = 3
2 + 4 = 6	2 + 8 = 10	2 + 9 = 11	2 + 6 = 8
2 + 7 = 9	2 + 10 = 12	2 + 6 = 8	2 + 4 = 6
2 + 4 = 6	2 + 7 = 9	2 + 2 = 4	2 + 7 = 9

DAY 04 ADDING DIGITS 0-6 SCORE: TIME:

3 + 2 = 5	3 + 10 = 13	3 + 7 = 10	3 + 8 = 11
3 + 1 = 4	3 + 7 = 10	3 + 2 = 5	3 + 10 = 13
3 + 5 = 8	3 + 10 = 13	3 + 6 = 9	3 + 8 = 11
3 + 2 = 5	3 + 8 = 11	3 + 5 = 8	3 + 10 = 13
3 + 3 = 6	3 + 6 = 9	3 + 3 = 6	3 + 6 = 9
3 + 6 = 9	3 + 9 = 12	3 + 5 = 8	3 + 4 = 7
3 + 6 = 9	3 + 1 = 4	3 + 0 = 3	3 + 6 = 9
3 + 0 = 3	3 + 6 = 9	3 + 7 = 10	3 + 10 = 13
3 + 6 = 9	3 + 0 = 3	3 + 1 = 4	3 + 2 = 5
3 + 8 = 11	3 + 9 = 12	3 + 6 = 9	3 + 2 = 5

ANSWER DAY 5-8

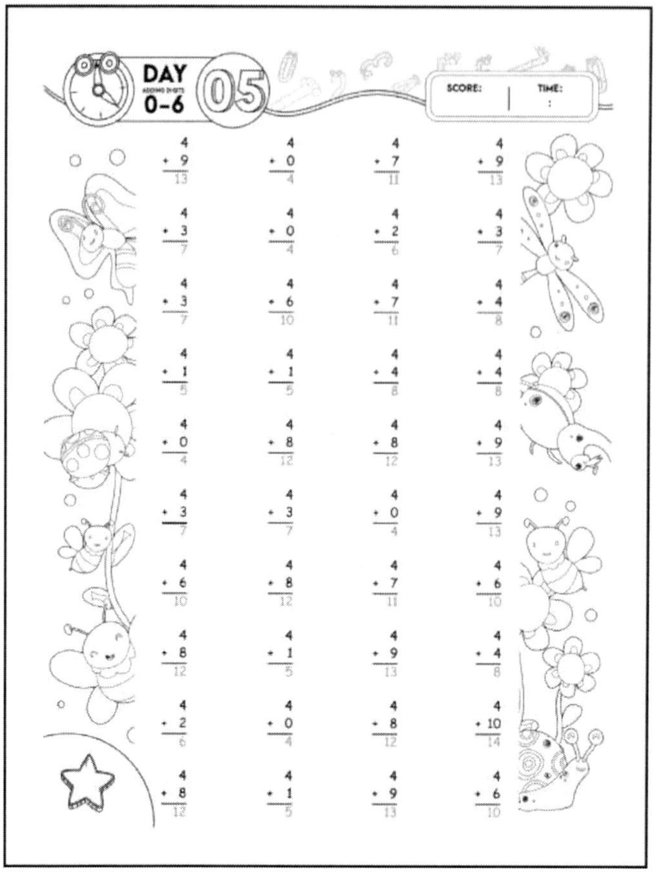

DAY 05
ADDING DIGITS 0-6
SCORE: TIME: :

4 + 9 = 13	4 + 0 = 4	4 + 7 = 11	4 + 9 = 13
4 + 3 = 7	4 + 0 = 4	4 + 2 = 6	4 + 3 = 7
4 + 3 = 7	4 + 6 = 10	4 + 7 = 11	4 + 4 = 8
4 + 1 = 5	4 + 1 = 5	4 + 4 = 8	4 + 4 = 8
4 + 0 = 4	4 + 8 = 12	4 + 8 = 12	4 + 9 = 13
4 + 3 = 7	4 + 3 = 7	4 + 0 = 4	4 + 9 = 13
4 + 6 = 10	4 + 8 = 12	4 + 7 = 11	4 + 6 = 10
4 + 8 = 12	4 + 1 = 5	4 + 9 = 13	4 + 4 = 8
4 + 2 = 6	4 + 0 = 4	4 + 8 = 12	4 + 10 = 14
4 + 8 = 12	4 + 1 = 5	4 + 9 = 13	4 + 6 = 10

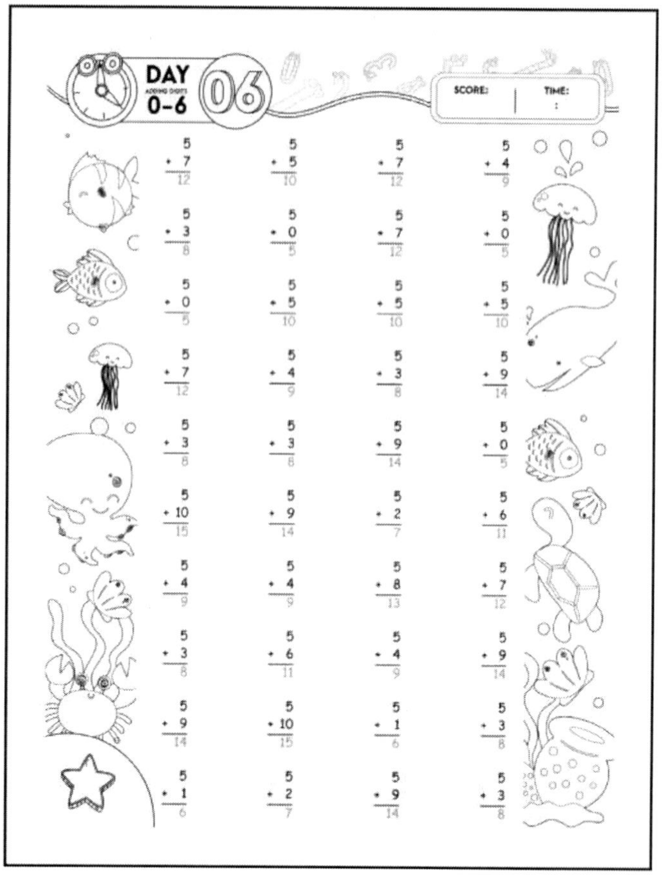

DAY 06
ADDING DIGITS 0-6
SCORE: TIME: :

5 + 7 = 12	5 + 5 = 10	5 + 7 = 12	5 + 4 = 9
5 + 3 = 8	5 + 0 = 5	5 + 7 = 12	5 + 0 = 5
5 + 0 = 5	5 + 5 = 10	5 + 5 = 10	5 + 5 = 10
5 + 7 = 12	5 + 4 = 9	5 + 3 = 8	5 + 9 = 14
5 + 3 = 8	5 + 3 = 8	5 + 9 = 14	5 + 0 = 5
5 + 10 = 15	5 + 9 = 14	5 + 2 = 7	5 + 6 = 11
5 + 4 = 9	5 + 4 = 9	5 + 8 = 13	5 + 7 = 12
5 + 3 = 8	5 + 6 = 11	5 + 4 = 9	5 + 9 = 14
5 + 9 = 14	5 + 10 = 15	5 + 1 = 6	5 + 3 = 8
5 + 1 = 6	5 + 2 = 7	5 + 9 = 14	5 + 3 = 8

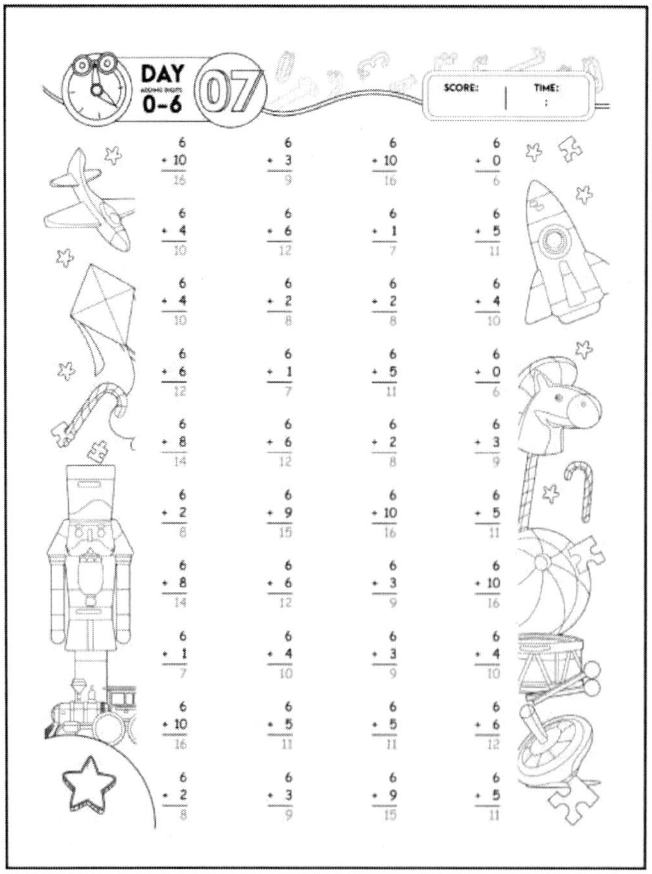

DAY 07
ADDING DIGITS 0-6
SCORE: TIME: :

6 + 10 = 16	6 + 3 = 9	6 + 10 = 16	6 + 0 = 6
6 + 4 = 10	6 + 6 = 12	6 + 1 = 7	6 + 5 = 11
6 + 4 = 10	6 + 2 = 8	6 + 2 = 8	6 + 4 = 10
6 + 6 = 12	6 + 1 = 7	6 + 5 = 11	6 + 0 = 6
6 + 8 = 14	6 + 6 = 12	6 + 2 = 8	6 + 3 = 9
6 + 2 = 8	6 + 9 = 15	6 + 10 = 16	6 + 5 = 11
6 + 8 = 14	6 + 6 = 12	6 + 3 = 9	6 + 10 = 16
6 + 1 = 7	6 + 4 = 10	6 + 3 = 9	6 + 4 = 10
6 + 10 = 16	6 + 5 = 11	6 + 5 = 11	6 + 6 = 12
6 + 2 = 8	6 + 3 = 9	6 + 9 = 15	6 + 5 = 11

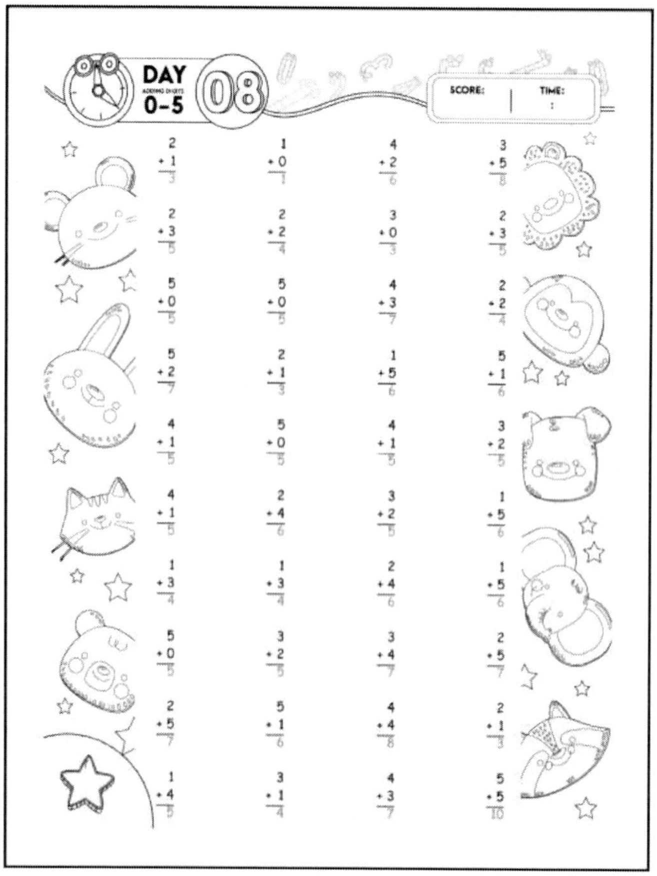

DAY 08
ADDING DIGITS 0-5
SCORE: TIME: :

2 + 1 = 3	1 + 0 = 1	4 + 2 = 6	3 + 5 = 8
2 + 3 = 5	2 + 2 = 4	3 + 0 = 3	2 + 3 = 5
5 + 0 = 5	5 + 0 = 5	4 + 3 = 7	2 + 2 = 4
5 + 2 = 7	2 + 1 = 3	1 + 5 = 6	5 + 1 = 6
4 + 1 = 5	5 + 0 = 5	4 + 1 = 5	3 + 2 = 5
4 + 1 = 5	2 + 4 = 6	3 + 2 = 5	1 + 5 = 6
1 + 3 = 4	1 + 3 = 4	2 + 4 = 6	1 + 5 = 6
5 + 0 = 5	3 + 2 = 5	3 + 4 = 7	2 + 5 = 7
2 + 5 = 7	5 + 1 = 6	4 + 4 = 8	2 + 1 = 3
1 + 4 = 5	3 + 1 = 4	4 + 3 = 7	5 + 5 = 10

ANSWER DAY 9-12

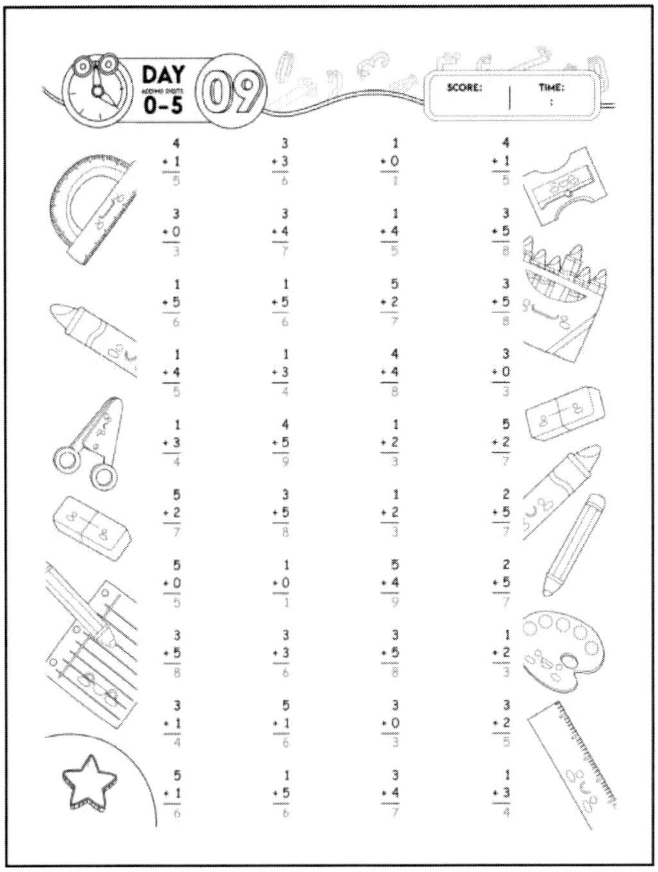

DAY 09 ADDING DIGITS 0-5

SCORE: | TIME: :

4 + 1 = 5	3 + 3 = 6	1 + 0 = 1	4 + 1 = 5
3 + 0 = 3	3 + 4 = 7	1 + 4 = 5	3 + 5 = 8
1 + 5 = 6	1 + 5 = 6	5 + 2 = 7	3 + 5 = 8
1 + 4 = 5	1 + 3 = 4	4 + 4 = 8	3 + 0 = 3
1 + 3 = 4	4 + 5 = 9	1 + 2 = 3	5 + 2 = 7
5 + 2 = 7	3 + 5 = 8	1 + 2 = 3	2 + 5 = 7
5 + 0 = 5	1 + 0 = 1	5 + 4 = 9	2 + 5 = 7
3 + 5 = 8	3 + 3 = 6	3 + 5 = 8	1 + 2 = 3
3 + 1 = 4	5 + 1 = 6	3 + 0 = 3	3 + 2 = 5
5 + 1 = 6	1 + 5 = 6	3 + 4 = 7	1 + 3 = 4

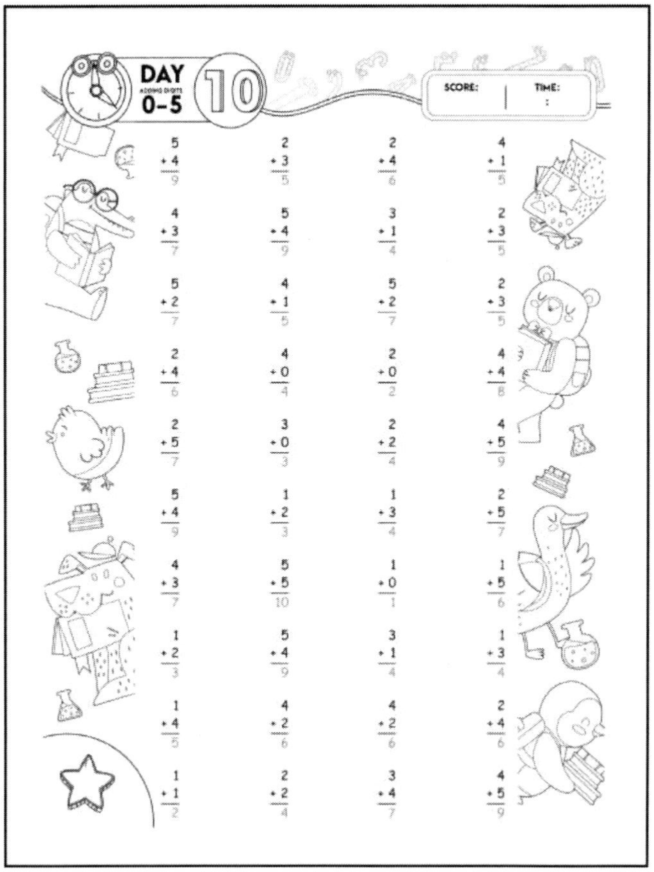

DAY 10 ADDING DIGITS 0-5

SCORE: | TIME: :

5 + 4 = 9	2 + 3 = 5	2 + 4 = 6	4 + 1 = 5
4 + 3 = 7	5 + 4 = 9	3 + 1 = 4	2 + 3 = 5
5 + 2 = 7	4 + 1 = 5	5 + 2 = 7	2 + 3 = 5
2 + 4 = 6	4 + 0 = 4	2 + 0 = 2	4 + 4 = 8
2 + 5 = 7	3 + 0 = 3	2 + 2 = 4	4 + 5 = 9
5 + 4 = 9	1 + 2 = 3	1 + 3 = 4	2 + 5 = 7
4 + 3 = 7	5 + 5 = 10	1 + 0 = 1	1 + 5 = 6
1 + 2 = 3	5 + 4 = 9	3 + 1 = 4	1 + 3 = 4
1 + 4 = 5	4 + 2 = 6	4 + 2 = 6	2 + 4 = 6
1 + 1 = 2	2 + 2 = 4	3 + 4 = 7	4 + 5 = 9

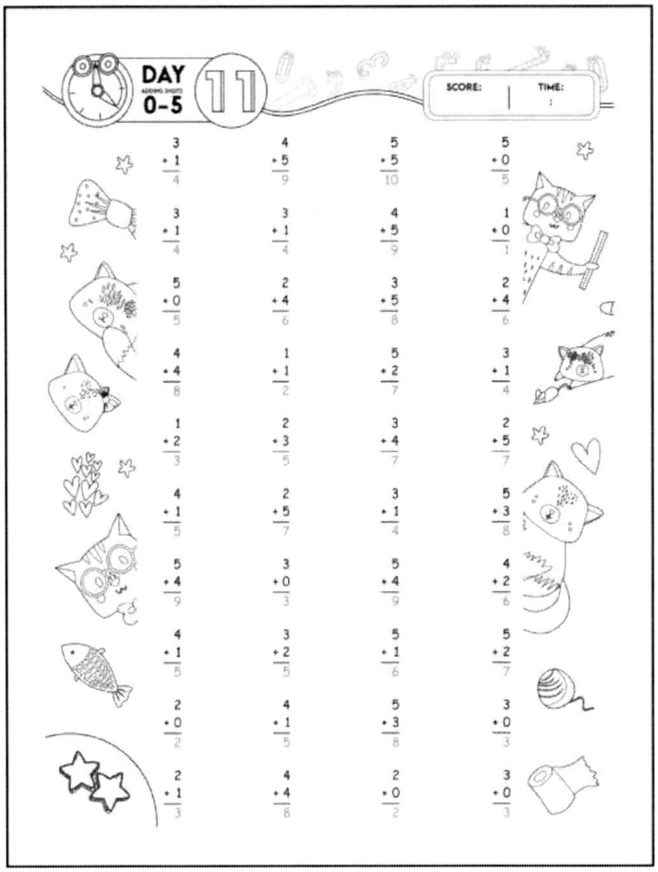

DAY 11 ADDING DIGITS 0-5

SCORE: | TIME: :

3 + 1 = 4	4 + 5 = 9	5 + 5 = 10	5 + 0 = 5
3 + 1 = 4	3 + 1 = 4	4 + 5 = 9	1 + 0 = 1
5 + 0 = 5	2 + 4 = 6	3 + 5 = 8	2 + 4 = 6
4 + 4 = 8	1 + 1 = 2	5 + 2 = 7	3 + 1 = 4
1 + 2 = 3	2 + 3 = 5	3 + 4 = 7	2 + 5 = 7
4 + 1 = 5	2 + 5 = 7	3 + 1 = 4	5 + 3 = 8
5 + 4 = 9	3 + 0 = 3	5 + 4 = 9	4 + 2 = 6
4 + 1 = 5	3 + 2 = 5	5 + 1 = 6	5 + 2 = 7
2 + 0 = 2	4 + 1 = 5	5 + 3 = 8	3 + 0 = 3
2 + 1 = 3	4 + 4 = 8	2 + 0 = 2	3 + 0 = 3

DAY 12 ADDING DIGITS 0-5

SCORE: | TIME: :

1 + 2 = 3	4 + 2 = 6	3 + 4 = 7	2 + 3 = 5
4 + 5 = 9	1 + 5 = 6	3 + 3 = 6	1 + 5 = 6
1 + 5 = 6	3 + 0 = 3	2 + 1 = 3	5 + 4 = 9
5 + 1 = 6	2 + 5 = 7	4 + 1 = 5	4 + 1 = 5
4 + 2 = 6	5 + 3 = 8	4 + 0 = 4	1 + 0 = 1
3 + 2 = 5	2 + 3 = 5	3 + 1 = 4	4 + 5 = 9
1 + 5 = 6	1 + 2 = 3	1 + 4 = 5	4 + 5 = 9
1 + 4 = 5	2 + 0 = 2	5 + 1 = 6	3 + 2 = 5
1 + 4 = 5	5 + 2 = 7	1 + 4 = 5	2 + 3 = 5
1 + 5 = 6	4 + 0 = 4	5 + 0 = 5	5 + 2 = 7

ANSWER DAY 13-16

DAY 13 — ADDING DIGITS 0-5

SCORE: | TIME: :

4 + 2 = 6	5 + 1 = 6	4 + 4 = 8	2 + 1 = 3
1 + 3 = 4	5 + 2 = 7	4 + 4 = 8	4 + 3 = 7
1 + 2 = 3	1 + 1 = 2	5 + 0 = 5	5 + 4 = 9
3 + 3 = 6	4 + 0 = 4	3 + 4 = 7	2 + 0 = 2
3 + 4 = 7	4 + 0 = 4	3 + 4 = 7	3 + 0 = 3
3 + 3 = 6	1 + 1 = 2	2 + 0 = 2	2 + 1 = 3
2 + 5 = 7	5 + 2 = 7	2 + 4 = 6	2 + 5 = 7
4 + 2 = 6	5 + 3 = 8	3 + 0 = 3	5 + 1 = 6
1 + 0 = 1	1 + 3 = 4	4 + 3 = 7	1 + 5 = 6
5 + 2 = 7	1 + 3 = 4	2 + 2 = 4	1 + 0 = 1

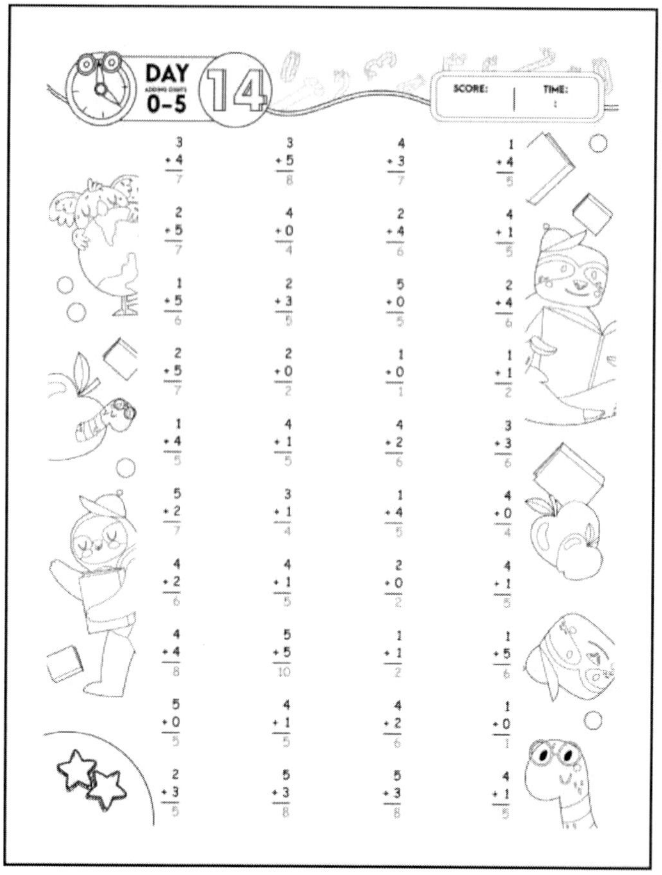

DAY 14 — ADDING DIGITS 0-5

SCORE: | TIME: :

3 + 4 = 7	3 + 5 = 8	4 + 3 = 7	1 + 4 = 5
2 + 5 = 7	4 + 0 = 4	2 + 4 = 6	4 + 1 = 5
1 + 5 = 6	2 + 3 = 5	5 + 0 = 5	2 + 4 = 6
2 + 5 = 7	2 + 0 = 2	1 + 0 = 1	1 + 1 = 2
1 + 4 = 5	4 + 1 = 5	4 + 2 = 6	3 + 3 = 6
5 + 2 = 7	3 + 1 = 4	1 + 4 = 5	4 + 0 = 4
4 + 2 = 6	4 + 1 = 5	2 + 0 = 2	4 + 1 = 5
4 + 4 = 8	5 + 5 = 10	1 + 1 = 2	1 + 5 = 6
5 + 0 = 5	4 + 1 = 5	4 + 2 = 6	1 + 0 = 1
2 + 3 = 5	5 + 3 = 8	5 + 3 = 8	4 + 1 = 5

DAY 15 — ADDING DIGITS 0-5

SCORE: | TIME: :

2 + 5 = 7	2 + 4 = 6	5 + 0 = 5	2 + 0 = 2
3 + 3 = 6	1 + 0 = 1	1 + 2 = 3	4 + 4 = 8
3 + 5 = 8	1 + 5 = 6	3 + 1 = 4	5 + 1 = 6
4 + 2 = 6	3 + 1 = 4	5 + 5 = 10	1 + 3 = 4
1 + 4 = 5	3 + 1 = 4	4 + 0 = 4	2 + 5 = 7
5 + 2 = 7	2 + 2 = 4	3 + 3 = 6	1 + 4 = 5
4 + 1 = 5	5 + 1 = 6	2 + 0 = 2	4 + 0 = 4
5 + 1 = 6	2 + 0 = 2	4 + 2 = 6	2 + 1 = 3
4 + 4 = 8	3 + 4 = 7	5 + 5 = 10	2 + 1 = 3
5 + 5 = 10	1 + 5 = 6	1 + 2 = 3	3 + 2 = 5

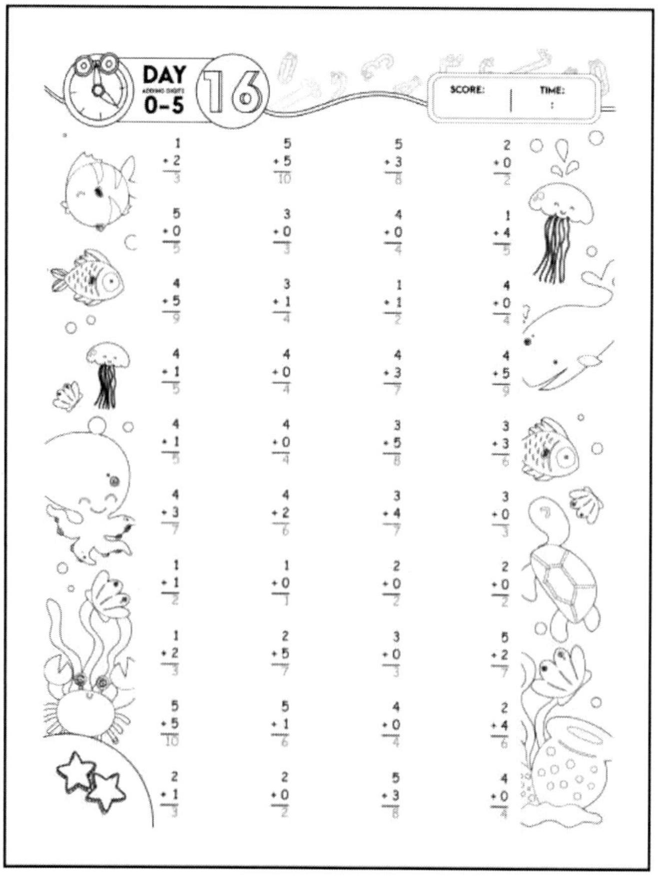

DAY 16 — ADDING DIGITS 0-5

SCORE: | TIME: :

1 + 2 = 3	5 + 5 = 10	5 + 3 = 8	2 + 0 = 2
5 + 0 = 5	3 + 0 = 3	4 + 0 = 4	1 + 4 = 5
4 + 5 = 9	3 + 1 = 4	1 + 1 = 2	4 + 0 = 4
4 + 1 = 5	4 + 0 = 4	4 + 3 = 7	4 + 5 = 9
4 + 1 = 5	4 + 0 = 4	3 + 5 = 8	3 + 3 = 6
4 + 3 = 7	4 + 2 = 6	3 + 4 = 7	3 + 0 = 3
1 + 1 = 2	1 + 0 = 1	2 + 0 = 2	2 + 0 = 2
1 + 2 = 3	2 + 5 = 7	3 + 0 = 3	5 + 2 = 7
5 + 5 = 10	5 + 1 = 6	4 + 0 = 4	2 + 4 = 6
2 + 1 = 3	2 + 0 = 2	5 + 3 = 8	4 + 0 = 4

ANSWER DAY 17-20

DAY 17

ADDING DIGITS 0-10

SCORE: | TIME: :

9 + 2 11	3 + 2 5	4 + 4 8	5 + 0 5
2 + 0 2	7 + 3 10	1 + 2 3	8 + 1 9
5 + 3 8	5 + 4 9	6 + 3 9	5 + 4 9
4 + 3 7	4 + 5 9	1 + 2 3	3 + 4 7
2 + 3 5	7 + 2 9	6 + 2 8	9 + 5 14
6 + 3 9	5 + 5 10	3 + 2 5	4 + 1 5
5 + 5 10	2 + 4 6	1 + 1 2	5 + 1 6
5 + 2 7	8 + 5 13	7 + 5 12	5 + 5 10
4 + 5 9	9 + 2 11	7 + 4 11	9 + 5 14
8 + 2 10	5 + 2 7	9 + 0 9	7 + 4 11

DAY 18

ADDING DIGITS 0-10

SCORE: | TIME: :

2 + 4 6	6 + 0 6	5 + 4 9	7 + 3 10
4 + 0 4	7 + 0 7	4 + 2 6	5 + 3 8
9 + 2 11	2 + 1 3	1 + 4 5	7 + 0 7
5 + 5 10	5 + 3 8	8 + 2 10	3 + 5 8
1 + 4 5	2 + 0 2	1 + 4 5	9 + 3 12
7 + 3 10	9 + 1 10	7 + 4 11	6 + 2 8
7 + 0 7	9 + 1 10	7 + 3 10	5 + 5 10
7 + 0 7	2 + 5 7	1 + 3 4	3 + 0 3
5 + 0 5	7 + 2 9	7 + 4 11	1 + 1 2
7 + 1 8	2 + 5 7	8 + 2 10	7 + 4 11

DAY 19

ADDING DIGITS 0-10

SCORE: | TIME: :

9 + 4 13	9 + 4 13	5 + 0 5	8 + 2 10
9 + 2 11	2 + 4 6	4 + 1 5	8 + 0 8
3 + 3 6	4 + 4 8	6 + 4 10	1 + 5 6
5 + 3 8	5 + 2 7	1 + 1 2	5 + 1 6
7 + 4 11	5 + 1 6	7 + 3 10	1 + 2 3
4 + 5 9	2 + 4 6	2 + 0 2	1 + 2 3
1 + 5 6	6 + 1 7	4 + 2 6	6 + 4 10
4 + 4 8	9 + 4 13	5 + 4 9	7 + 1 8
6 + 2 8	7 + 5 12	3 + 5 8	4 + 1 5
2 + 4 6	3 + 1 4	2 + 2 4	5 + 5 10

DAY 20

ADDING DIGITS 0-10

SCORE: | TIME: :

5 + 0 5	7 + 1 8	3 + 5 8	6 + 1 7
8 + 1 9	3 + 2 5	7 + 5 12	8 + 1 9
6 + 3 9	6 + 2 8	9 + 5 14	3 + 5 8
8 + 0 8	8 + 3 11	4 + 0 4	9 + 0 9
8 + 1 9	2 + 3 5	3 + 0 3	1 + 5 6
4 + 2 6	5 + 4 9	2 + 5 7	7 + 5 12
1 + 0 1	2 + 2 4	3 + 1 4	1 + 3 4
3 + 4 7	7 + 1 8	9 + 4 13	3 + 1 4
5 + 0 5	7 + 0 7	8 + 3 11	1 + 1 2
4 + 4 8	2 + 0 2	5 + 5 10	4 + 0 4

ANSWER DAY 21-24

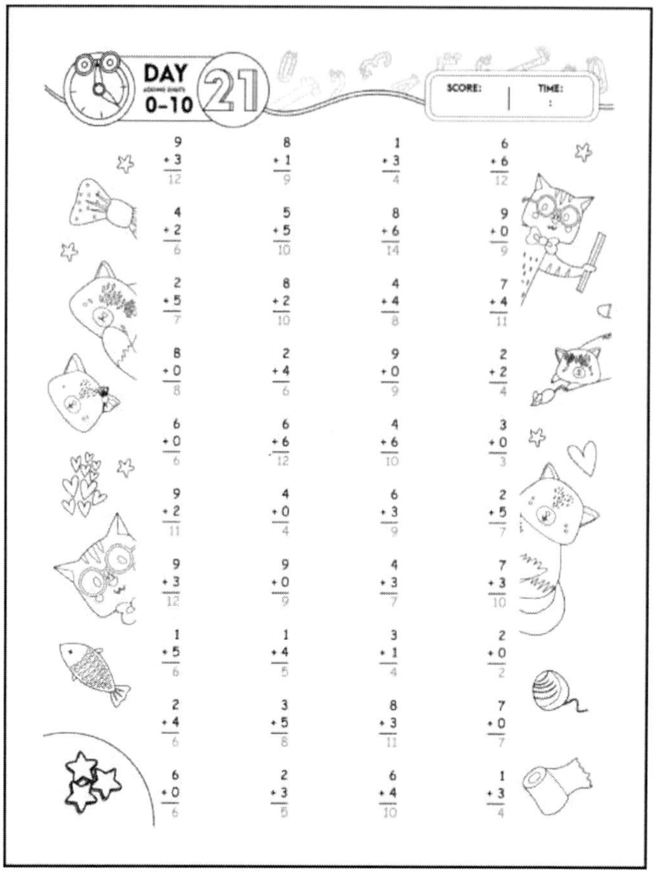

DAY 21
ADDING DIGITS
0-10
SCORE: | TIME: :

9 + 3 12	8 + 1 9	1 + 3 4	6 + 6 12
4 + 2 6	5 + 5 10	8 + 6 14	9 + 0 9
2 + 5 7	8 + 2 10	4 + 4 8	7 + 4 11
8 + 0 8	2 + 4 6	9 + 0 9	2 + 2 4
6 + 0 6	6 + 6 12	4 + 6 10	3 + 0 3
9 + 2 11	4 + 0 4	6 + 3 9	2 + 5 7
9 + 3 12	9 + 0 9	4 + 3 7	7 + 3 10
1 + 5 6	1 + 4 5	3 + 1 4	2 + 0 2
2 + 4 6	3 + 5 8	8 + 3 11	7 + 0 7
6 + 0 6	2 + 3 5	6 + 4 10	1 + 3 4

DAY 22
ADDING DIGITS
0-10
SCORE: | TIME: :

2 + 6 8	5 + 6 11	9 + 0 9	3 + 2 5
5 + 0 5	4 + 4 8	3 + 2 5	1 + 1 2
6 + 4 10	3 + 6 9	4 + 1 5	2 + 6 8
7 + 5 12	9 + 2 11	5 + 1 6	5 + 3 8
8 + 6 14	2 + 5 7	6 + 4 10	7 + 4 11
8 + 6 14	3 + 2 5	2 + 5 7	2 + 3 5
3 + 6 9	5 + 3 8	2 + 1 3	8 + 4 12
1 + 0 1	1 + 4 5	1 + 4 5	5 + 6 11
6 + 3 9	4 + 3 7	2 + 0 2	6 + 4 10
9 + 5 14	2 + 1 3	6 + 4 10	1 + 6 7

DAY 23
ADDING DIGITS
0-10
SCORE: | TIME: :

2 + 3 5	9 + 5 14	7 + 3 10	6 + 0 6
9 + 0 9	8 + 0 8	5 + 0 5	7 + 1 8
3 + 3 6	2 + 5 7	8 + 3 11	9 + 3 12
8 + 0 8	5 + 2 7	4 + 6 10	3 + 4 7
3 + 0 3	8 + 4 12	8 + 3 11	1 + 5 6
3 + 2 5	5 + 1 6	4 + 0 4	7 + 4 11
5 + 5 10	2 + 6 8	7 + 0 7	5 + 0 5
3 + 3 6	2 + 0 2	5 + 2 7	8 + 1 9
6 + 5 11	7 + 1 8	8 + 3 11	8 + 6 14
2 + 5 7	1 + 2 3	8 + 1 9	4 + 5 9

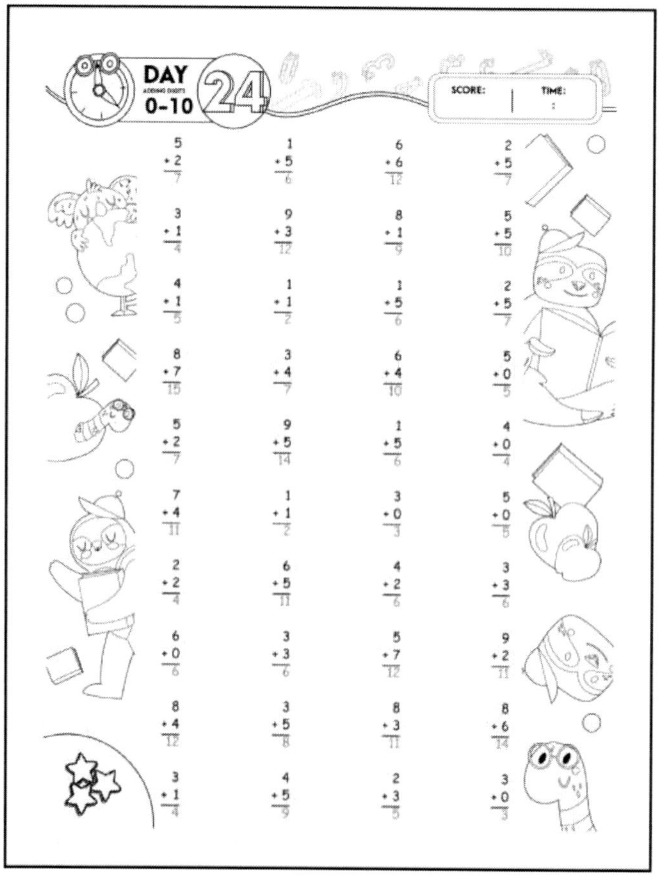

DAY 24
ADDING DIGITS
0-10
SCORE: | TIME: :

5 + 2 7	1 + 5 6	6 + 6 12	2 + 5 7
3 + 1 4	9 + 3 12	8 + 1 9	5 + 5 10
4 + 1 5	1 + 1 2	1 + 5 6	2 + 5 7
8 + 7 15	3 + 4 7	6 + 4 10	5 + 0 5
5 + 2 7	9 + 5 14	1 + 5 6	4 + 0 4
7 + 4 11	1 + 1 2	3 + 0 3	5 + 0 5
2 + 2 4	6 + 5 11	4 + 2 6	3 + 3 6
6 + 0 6	3 + 3 6	5 + 7 12	9 + 2 11
8 + 4 12	3 + 5 8	8 + 3 11	8 + 6 14
3 + 1 4	4 + 5 9	2 + 3 5	3 + 0 3

ANSWER DAY 25-28

DAY 25

ADDING DIGITS 0-10

SCORE: | TIME: :

3 + 0 3	7 + 5 12	9 + 4 13	9 + 2 11
7 + 6 13	4 + 2 6	3 + 3 6	4 + 6 10
3 + 7 10	6 + 7 13	5 + 3 8	8 + 0 8
3 + 0 3	5 + 3 8	2 + 4 6	8 + 7 15
4 + 0 4	3 + 4 7	3 + 7 10	8 + 6 14
5 + 1 6	1 + 1 2	6 + 6 12	8 + 2 10
6 + 1 7	6 + 6 12	9 + 6 15	6 + 5 11
5 + 6 11	3 + 5 8	4 + 3 7	1 + 7 8
3 + 0 3	2 + 0 2	3 + 7 10	5 + 6 11
3 + 5 8	5 + 5 10	9 + 4 13	3 + 3 6

DAY 26

ADDING DIGITS 0-10

SCORE: | TIME: :

7 + 3 10	3 + 6 9	2 + 6 8	1 + 4 5
9 + 0 9	7 + 3 10	6 + 0 6	6 + 1 7
9 + 3 12	5 + 0 5	4 + 2 6	3 + 4 7
5 + 3 8	6 + 2 8	2 + 6 8	2 + 1 3
2 + 1 3	1 + 5 6	7 + 5 12	4 + 0 4
4 + 1 5	9 + 2 11	9 + 0 9	1 + 0 1
3 + 7 10	7 + 3 10	7 + 6 13	9 + 6 15
9 + 6 15	8 + 5 13	3 + 2 5	6 + 0 6
6 + 5 11	5 + 2 7	3 + 4 7	8 + 1 9
8 + 6 14	8 + 0 8	5 + 0 5	1 + 1 2

DAY 27

ADDING DIGITS 0-10

SCORE: | TIME: :

1 + 2 3	3 + 3 6	6 + 7 13	4 + 3 7
2 + 4 6	8 + 2 10	6 + 6 12	8 + 7 15
3 + 5 8	5 + 6 11	9 + 3 12	2 + 5 7
7 + 7 14	9 + 0 9	7 + 2 9	4 + 5 9
1 + 6 7	2 + 4 6	7 + 8 15	8 + 0 8
6 + 7 13	3 + 5 8	8 + 6 14	8 + 0 8
9 + 8 17	6 + 1 7	1 + 6 7	4 + 4 8
8 + 5 13	1 + 6 7	7 + 3 10	8 + 7 15
2 + 7 9	6 + 1 7	1 + 7 8	5 + 2 7
9 + 1 10	6 + 6 12	8 + 4 12	6 + 7 13

DAY 28

ADDING DIGITS 0-10

SCORE: | TIME: :

4 + 3 7	5 + 3 8	5 + 8 13	4 + 7 11
3 + 3 6	7 + 0 7	8 + 8 16	4 + 1 5
3 + 6 9	9 + 2 11	8 + 1 9	7 + 0 7
6 + 5 11	2 + 3 5	4 + 0 4	5 + 4 9
1 + 3 4	8 + 2 10	8 + 8 16	9 + 7 16
3 + 5 8	1 + 7 8	8 + 4 12	9 + 7 16
7 + 2 9	4 + 2 6	7 + 6 13	9 + 8 17
5 + 4 9	9 + 4 13	8 + 7 15	2 + 5 7
4 + 2 6	8 + 5 13	7 + 4 11	7 + 6 13
9 + 5 14	1 + 6 7	1 + 7 8	5 + 5 10

ANSWER DAY 29-32

DAY 29
ADDING DIGITS 0-10

SCORE: | TIME: :

4 + 6 = 10	4 + 6 = 10	6 + 5 = 11	7 + 4 = 11
7 + 1 = 8	8 + 7 = 15	3 + 2 = 5	2 + 1 = 3
7 + 4 = 11	8 + 3 = 11	1 + 0 = 1	1 + 6 = 7
9 + 2 = 11	4 + 0 = 4	4 + 2 = 6	8 + 8 = 16
2 + 0 = 2	4 + 1 = 5	7 + 1 = 8	9 + 2 = 11
9 + 1 = 10	4 + 3 = 7	2 + 2 = 4	6 + 7 = 13
4 + 6 = 10	2 + 4 = 6	7 + 2 = 9	7 + 6 = 13
7 + 0 = 7	1 + 7 = 8	3 + 5 = 8	6 + 0 = 6
2 + 0 = 2	8 + 4 = 12	5 + 4 = 9	3 + 5 = 8
1 + 2 = 3	6 + 3 = 9	2 + 2 = 4	5 + 4 = 9

DAY 30
ADDING DIGITS 0-10

SCORE: | TIME: :

6 + 4 = 10	9 + 4 = 13	2 + 2 = 4	4 + 5 = 9
8 + 6 = 14	5 + 4 = 9	5 + 2 = 7	3 + 1 = 4
9 + 0 = 9	2 + 7 = 9	1 + 5 = 6	4 + 3 = 7
3 + 5 = 8	7 + 1 = 8	2 + 4 = 6	2 + 2 = 4
8 + 3 = 11	4 + 4 = 8	7 + 4 = 11	9 + 3 = 12
1 + 8 = 9	7 + 0 = 7	5 + 3 = 8	1 + 7 = 8
6 + 6 = 12	5 + 6 = 11	6 + 0 = 6	5 + 6 = 11
2 + 0 = 2	4 + 5 = 9	6 + 5 = 11	7 + 3 = 10
8 + 3 = 11	1 + 1 = 2	9 + 3 = 12	3 + 6 = 9
9 + 0 = 9	5 + 3 = 8	1 + 4 = 5	7 + 3 = 10

DAY 31
ADDING DIGITS 0-10

SCORE: | TIME: :

6 + 5 = 11	6 + 3 = 9	2 + 8 = 10	6 + 4 = 10
6 + 9 = 15	4 + 2 = 6	9 + 6 = 15	2 + 4 = 6
2 + 6 = 8	3 + 2 = 5	5 + 5 = 10	3 + 2 = 5
2 + 7 = 9	7 + 9 = 16	9 + 5 = 14	2 + 1 = 3
6 + 0 = 6	3 + 4 = 7	7 + 0 = 7	2 + 3 = 5
6 + 4 = 10	1 + 7 = 8	6 + 2 = 8	6 + 1 = 7
2 + 9 = 11	1 + 8 = 9	1 + 2 = 3	2 + 2 = 4
4 + 5 = 9	2 + 5 = 7	8 + 6 = 14	4 + 0 = 4
6 + 5 = 11	3 + 8 = 11	8 + 9 = 17	8 + 6 = 14
8 + 3 = 11	7 + 0 = 7	1 + 4 = 5	4 + 0 = 4

DAY 32
ADDING DIGITS 0-10

SCORE: | TIME: :

2 + 7 = 9	6 + 1 = 7	7 + 8 = 15	5 + 1 = 6
5 + 7 = 12	8 + 3 = 11	6 + 6 = 12	1 + 7 = 8
3 + 2 = 5	3 + 3 = 6	1 + 3 = 4	9 + 8 = 17
6 + 6 = 12	2 + 1 = 3	6 + 8 = 14	6 + 3 = 9
9 + 1 = 10	1 + 2 = 3	2 + 7 = 9	5 + 7 = 12
5 + 2 = 7	7 + 5 = 12	7 + 3 = 10	9 + 4 = 13
2 + 2 = 4	3 + 0 = 3	4 + 5 = 9	2 + 7 = 9
2 + 0 = 2	2 + 3 = 5	7 + 0 = 7	6 + 9 = 15
3 + 6 = 9	4 + 3 = 7	4 + 9 = 13	4 + 0 = 4
3 + 2 = 5	4 + 7 = 11	1 + 4 = 5	8 + 5 = 13

ANSWER DAY 33-36

DAY 33

ADDING DIGITS 0–10

SCORE: | TIME: :

9 + 1 = 10	9 + 3 = 12	9 + 7 = 16	9 + 7 = 16
4 + 1 = 5	5 + 5 = 10	5 + 4 = 9	4 + 3 = 7
1 + 5 = 6	4 + 1 = 5	8 + 9 = 17	3 + 5 = 8
6 + 6 = 12	3 + 9 = 12	8 + 6 = 14	2 + 1 = 3
9 + 0 = 9	1 + 0 = 1	8 + 8 = 16	4 + 1 = 5
9 + 3 = 12	9 + 1 = 10	9 + 7 = 16	3 + 4 = 7
9 + 2 = 11	2 + 9 = 11	5 + 8 = 13	2 + 5 = 7
4 + 7 = 11	1 + 3 = 4	9 + 9 = 18	6 + 4 = 10
4 + 6 = 10	1 + 5 = 6	5 + 2 = 7	9 + 5 = 14
8 + 1 = 9	4 + 7 = 11	8 + 9 = 17	7 + 0 = 7

DAY 34

ADDING DIGITS 0–10

SCORE: | TIME: :

1 + 5 = 6	8 + 6 = 14	3 + 5 = 8	4 + 3 = 7
8 + 3 = 11	8 + 8 = 16	9 + 3 = 12	5 + 0 = 5
1 + 0 = 1	3 + 2 = 5	9 + 5 = 14	4 + 4 = 8
8 + 0 = 8	7 + 4 = 11	4 + 9 = 13	8 + 7 = 15
4 + 5 = 9	9 + 6 = 15	7 + 5 = 12	6 + 9 = 15
5 + 2 = 7	4 + 9 = 13	2 + 6 = 8	7 + 6 = 13
9 + 0 = 9	6 + 8 = 14	5 + 7 = 12	5 + 9 = 14
8 + 9 = 17	4 + 4 = 8	7 + 9 = 16	1 + 3 = 4
7 + 5 = 12	2 + 7 = 9	7 + 3 = 10	2 + 5 = 7
6 + 8 = 14	9 + 2 = 11	5 + 7 = 12	5 + 2 = 7

DAY 35

ADDING DIGITS 0–10

SCORE: | TIME: :

3 + 0 = 3	5 + 2 = 7	5 + 10 = 15	5 + 10 = 15
2 + 7 = 9	9 + 3 = 12	6 + 10 = 16	7 + 1 = 8
7 + 1 = 8	1 + 1 = 2	8 + 5 = 13	6 + 10 = 16
3 + 6 = 9	5 + 0 = 5	6 + 6 = 12	5 + 3 = 8
2 + 6 = 8	6 + 10 = 16	4 + 9 = 13	8 + 5 = 13
3 + 6 = 9	6 + 0 = 6	7 + 1 = 8	7 + 4 = 11
1 + 1 = 2	3 + 8 = 11	1 + 2 = 3	3 + 10 = 13
4 + 7 = 11	4 + 6 = 10	6 + 6 = 12	1 + 10 = 11
6 + 10 = 16	1 + 1 = 2	8 + 5 = 13	1 + 1 = 2
5 + 7 = 12	1 + 10 = 11	5 + 1 = 6	4 + 7 = 11

DAY 36

ADDING DIGITS 0–10

SCORE: | TIME: :

7 + 6 = 13	3 + 3 = 6	9 + 1 = 10	7 + 7 = 14
4 + 7 = 11	2 + 1 = 3	8 + 0 = 8	3 + 9 = 12
8 + 7 = 15	9 + 2 = 11	2 + 9 = 11	9 + 5 = 14
6 + 9 = 15	4 + 1 = 5	6 + 1 = 7	3 + 8 = 11
2 + 1 = 3	8 + 7 = 15	4 + 5 = 9	5 + 3 = 8
5 + 0 = 5	2 + 7 = 9	4 + 10 = 14	1 + 6 = 7
6 + 6 = 12	8 + 5 = 13	1 + 6 = 7	5 + 1 = 6
6 + 6 = 12	9 + 3 = 12	3 + 5 = 8	3 + 7 = 10
9 + 2 = 11	4 + 10 = 14	5 + 4 = 9	2 + 8 = 10
8 + 6 = 14	6 + 6 = 12	4 + 1 = 5	2 + 4 = 6

ANSWER DAY 37-40

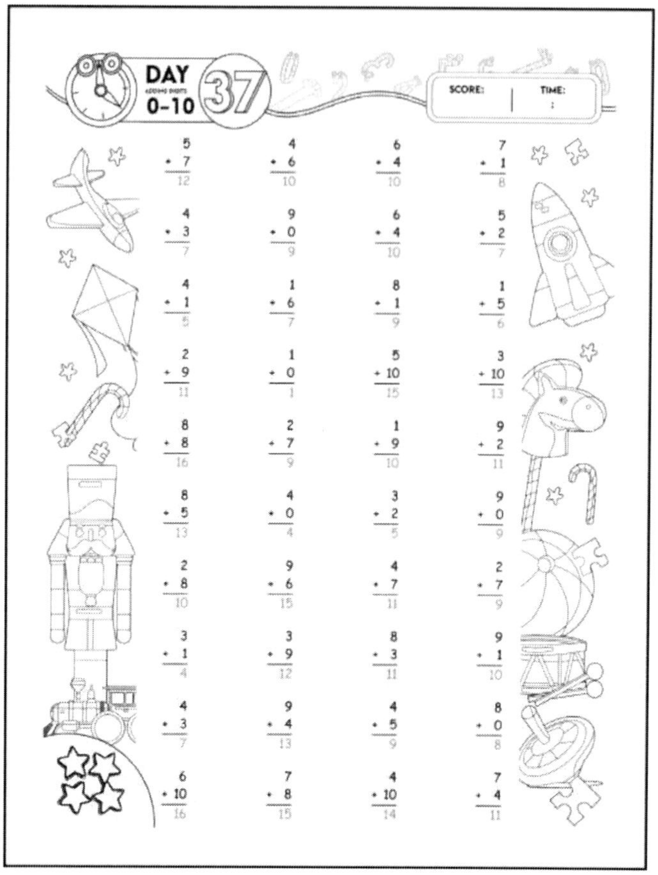

DAY 37 — ADDING DIGITS 0–10

SCORE: | TIME: :

5 + 7 = 12	4 + 6 = 10	6 + 4 = 10	7 + 1 = 8
4 + 3 = 7	9 + 0 = 9	6 + 4 = 10	5 + 2 = 7
4 + 1 = 5	1 + 6 = 7	8 + 1 = 9	1 + 5 = 6
2 + 9 = 11	1 + 0 = 1	5 + 10 = 15	3 + 10 = 13
8 + 8 = 16	2 + 7 = 9	1 + 9 = 10	9 + 2 = 11
8 + 5 = 13	4 + 0 = 4	3 + 2 = 5	9 + 0 = 9
2 + 8 = 10	9 + 6 = 15	4 + 7 = 11	2 + 7 = 9
3 + 1 = 4	3 + 9 = 12	8 + 3 = 11	9 + 1 = 10
4 + 3 = 7	9 + 4 = 13	4 + 5 = 9	8 + 0 = 8
6 + 10 = 16	7 + 8 = 15	4 + 10 = 14	7 + 4 = 11

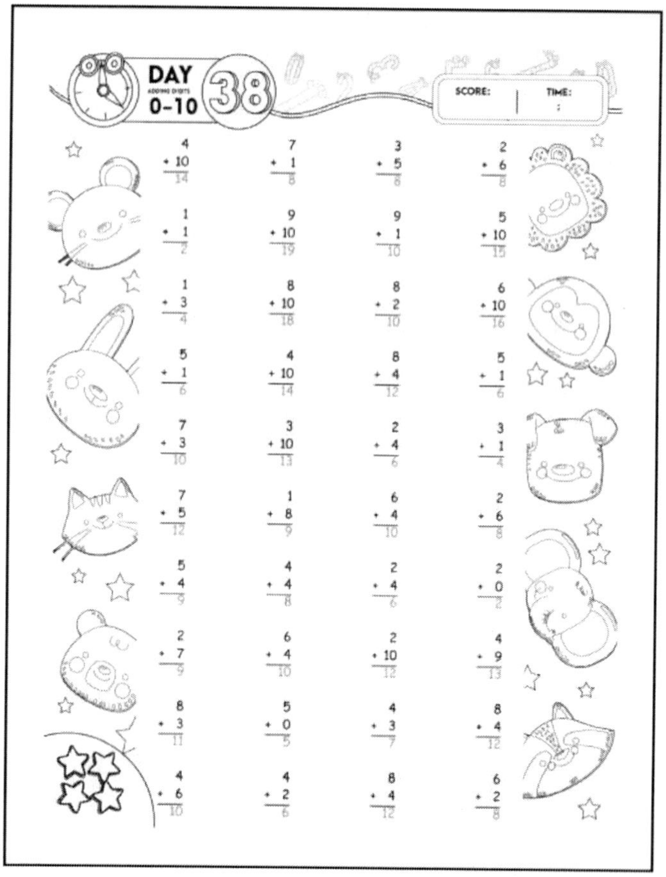

DAY 38 — ADDING DIGITS 0–10

SCORE: | TIME: :

4 + 10 = 14	7 + 1 = 8	3 + 5 = 8	2 + 6 = 8
1 + 1 = 2	9 + 10 = 19	9 + 1 = 10	5 + 10 = 15
1 + 3 = 4	8 + 10 = 18	8 + 2 = 10	6 + 10 = 16
5 + 1 = 6	4 + 10 = 14	8 + 4 = 12	5 + 1 = 6
7 + 3 = 10	3 + 10 = 13	2 + 4 = 6	3 + 1 = 4
7 + 5 = 12	1 + 8 = 9	6 + 4 = 10	2 + 6 = 8
5 + 4 = 9	4 + 4 = 8	2 + 4 = 6	2 + 0 = 2
2 + 7 = 9	6 + 4 = 10	2 + 10 = 12	4 + 9 = 13
8 + 3 = 11	5 + 0 = 5	4 + 3 = 7	8 + 4 = 12
4 + 6 = 10	4 + 2 = 6	8 + 4 = 12	6 + 2 = 8

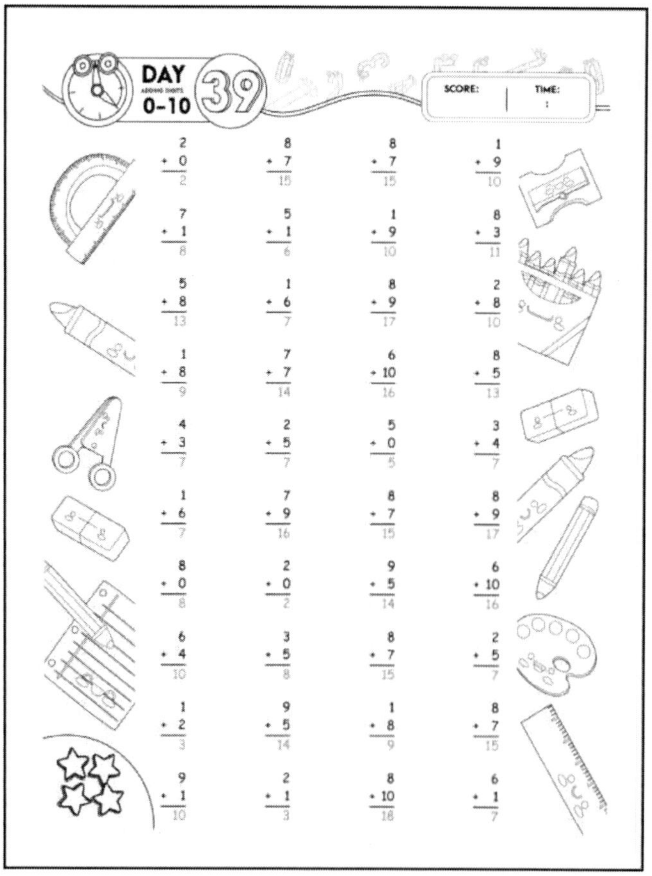

DAY 39 — ADDING DIGITS 0–10

SCORE: | TIME: :

2 + 0 = 2	8 + 7 = 15	8 + 7 = 15	1 + 9 = 10
7 + 1 = 8	5 + 1 = 6	1 + 9 = 10	8 + 3 = 11
5 + 8 = 13	1 + 6 = 7	8 + 9 = 17	2 + 8 = 10
1 + 8 = 9	7 + 7 = 14	6 + 10 = 16	8 + 5 = 13
4 + 3 = 7	2 + 5 = 7	5 + 0 = 5	3 + 4 = 7
1 + 6 = 7	7 + 9 = 16	8 + 7 = 15	8 + 9 = 17
8 + 0 = 8	2 + 0 = 2	9 + 5 = 14	6 + 10 = 16
6 + 4 = 10	3 + 5 = 8	8 + 7 = 15	2 + 5 = 7
1 + 2 = 3	9 + 5 = 14	1 + 8 = 9	8 + 7 = 15
9 + 1 = 10	2 + 1 = 3	8 + 10 = 18	6 + 1 = 7

DAY 40 — ADDING DIGITS 0–10

SCORE: | TIME: :

9 + 1 = 10	5 + 6 = 11	4 + 5 = 9	6 + 4 = 10
3 + 1 = 4	3 + 7 = 10	2 + 5 = 7	6 + 10 = 16
5 + 0 = 5	7 + 0 = 7	8 + 2 = 10	3 + 1 = 4
7 + 10 = 17	5 + 5 = 10	8 + 7 = 15	5 + 7 = 12
2 + 8 = 10	9 + 5 = 14	8 + 0 = 8	2 + 7 = 9
5 + 0 = 5	3 + 5 = 8	9 + 3 = 12	1 + 8 = 9
4 + 2 = 6	6 + 6 = 12	7 + 4 = 11	4 + 4 = 8
8 + 9 = 17	4 + 5 = 9	8 + 6 = 14	3 + 10 = 13
3 + 2 = 5	8 + 8 = 16	4 + 4 = 8	8 + 9 = 17
6 + 4 = 10	9 + 8 = 17	5 + 4 = 9	6 + 3 = 9

ANSWER DAY 41-44

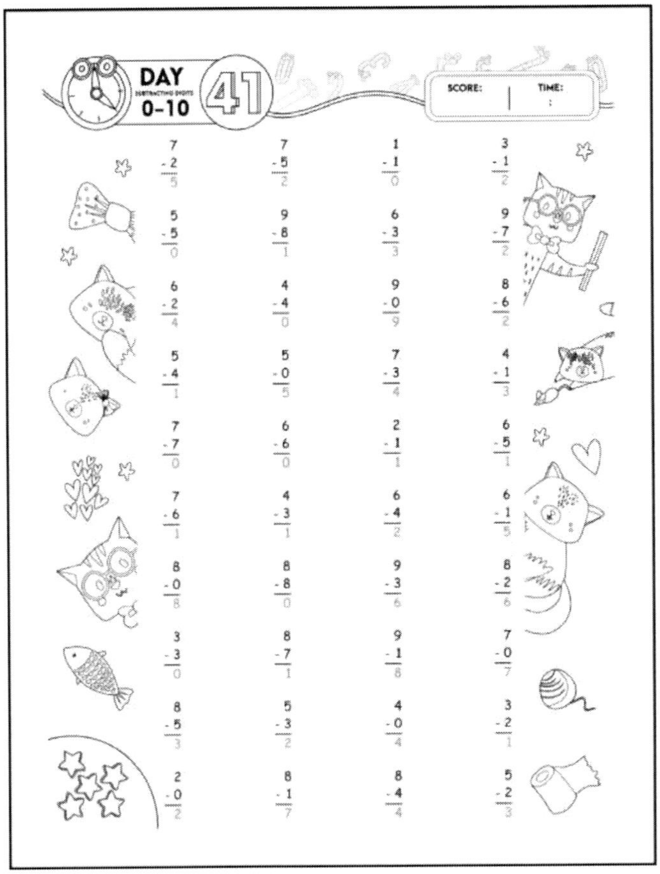

DAY 41
SUBTRACTING DIGITS
0-10

SCORE: | TIME: :

7 - 2 = 5	7 - 5 = 2	1 - 1 = 0	3 - 1 = 2
5 - 5 = 0	9 - 8 = 1	6 - 3 = 3	9 - 7 = 2
6 - 2 = 4	4 - 4 = 0	9 - 0 = 9	8 - 6 = 2
5 - 4 = 1	5 - 0 = 5	7 - 3 = 4	4 - 1 = 3
7 - 7 = 0	6 - 6 = 0	2 - 1 = 1	6 - 5 = 1
7 - 6 = 1	4 - 3 = 1	6 - 4 = 2	6 - 1 = 5
8 - 0 = 8	8 - 8 = 0	9 - 3 = 6	8 - 2 = 6
3 - 3 = 0	8 - 7 = 1	9 - 1 = 8	7 - 0 = 7
8 - 5 = 3	5 - 3 = 2	4 - 0 = 4	3 - 2 = 1
2 - 0 = 2	8 - 1 = 7	8 - 4 = 4	5 - 2 = 3

DAY 42
SUBTRACTING DIGITS
0-10

SCORE: | TIME: :

7 - 0 = 7	1 - 0 = 1	8 - 3 = 5	6 - 4 = 2
7 - 2 = 5	6 - 5 = 1	5 - 0 = 5	4 - 3 = 1
9 - 9 = 0	9 - 6 = 3	5 - 4 = 1	2 - 1 = 1
8 - 0 = 8	9 - 7 = 2	7 - 6 = 1	2 - 2 = 0
4 - 4 = 0	9 - 3 = 6	4 - 1 = 3	2 - 0 = 2
4 - 0 = 4	6 - 0 = 6	3 - 1 = 2	5 - 1 = 4
8 - 5 = 3	5 - 5 = 0	3 - 2 = 1	8 - 2 = 6
9 - 5 = 4	6 - 2 = 4	8 - 7 = 1	6 - 6 = 0
3 - 3 = 0	9 - 2 = 7	5 - 2 = 3	7 - 4 = 3
9 - 1 = 8	9 - 0 = 9	8 - 1 = 7	6 - 3 = 3

DAY 43
SUBTRACTING DIGITS
0-10

SCORE: | TIME: :

5 - 4 = 1	8 - 7 = 1	2 - 2 = 0	7 - 2 = 5
9 - 2 = 7	2 - 0 = 2	1 - 0 = 1	6 - 1 = 5
8 - 2 = 6	2 - 1 = 1	9 - 5 = 4	9 - 3 = 6
4 - 4 = 0	4 - 1 = 3	9 - 0 = 9	9 - 9 = 0
5 - 5 = 0	6 - 4 = 2	7 - 0 = 7	7 - 4 = 3
4 - 2 = 2	6 - 3 = 3	8 - 8 = 0	7 - 6 = 1
8 - 3 = 5	4 - 0 = 4	7 - 7 = 0	8 - 0 = 8
1 - 1 = 0	4 - 3 = 1	5 - 2 = 3	7 - 3 = 4
7 - 5 = 2	5 - 3 = 2	5 - 0 = 5	6 - 2 = 4
3 - 3 = 0	9 - 6 = 3	9 - 1 = 8	6 - 5 = 1

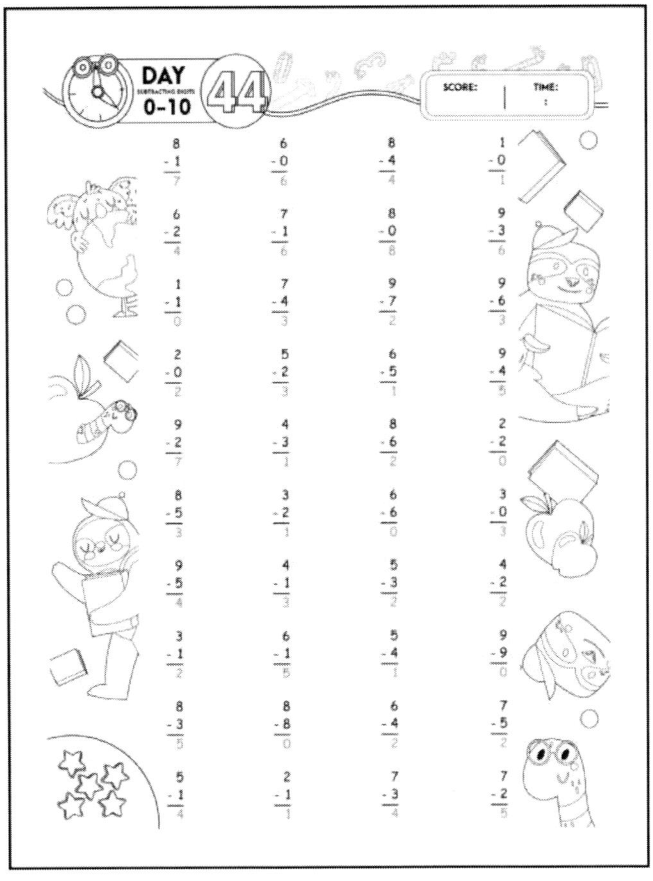

DAY 44
SUBTRACTING DIGITS
0-10

SCORE: | TIME: :

8 - 1 = 7	6 - 0 = 6	8 - 4 = 4	1 - 0 = 1
6 - 2 = 4	7 - 1 = 6	8 - 0 = 8	9 - 3 = 6
1 - 1 = 0	7 - 4 = 3	9 - 7 = 2	9 - 6 = 3
2 - 0 = 2	5 - 2 = 3	6 - 5 = 1	9 - 4 = 5
9 - 2 = 7	4 - 3 = 1	8 - 6 = 2	2 - 2 = 0
8 - 5 = 3	3 - 2 = 1	6 - 6 = 0	3 - 0 = 3
9 - 5 = 4	4 - 1 = 3	5 - 3 = 2	4 - 2 = 2
3 - 1 = 2	6 - 1 = 5	5 - 4 = 1	9 - 9 = 0
8 - 3 = 5	8 - 8 = 0	6 - 4 = 2	7 - 5 = 2
5 - 1 = 4	2 - 1 = 1	7 - 3 = 4	7 - 2 = 5

ANSWER DAY 45-48

DAY 45 SUBTRACTING DIGITS 0–10

SCORE: | TIME: :

6 - 0 = 6	4 - 0 = 4	3 - 3 = 0	4 - 4 = 0
8 - 7 = 1	5 - 0 = 5	5 - 4 = 1	4 - 1 = 3
3 - 2 = 1	9 - 4 = 5	9 - 6 = 3	2 - 2 = 0
9 - 0 = 9	7 - 4 = 3	9 - 3 = 6	6 - 2 = 4
5 - 5 = 0	8 - 3 = 5	8 - 1 = 7	8 - 8 = 0
4 - 3 = 1	6 - 1 = 5	9 - 7 = 2	8 - 5 = 3
6 - 6 = 0	7 - 7 = 0	7 - 2 = 5	6 - 5 = 1
7 - 6 = 1	7 - 0 = 7	9 - 8 = 1	1 - 0 = 1
9 - 2 = 7	2 - 1 = 1	7 - 3 = 4	8 - 0 = 8
1 - 1 = 0	6 - 4 = 2	3 - 1 = 2	8 - 6 = 2

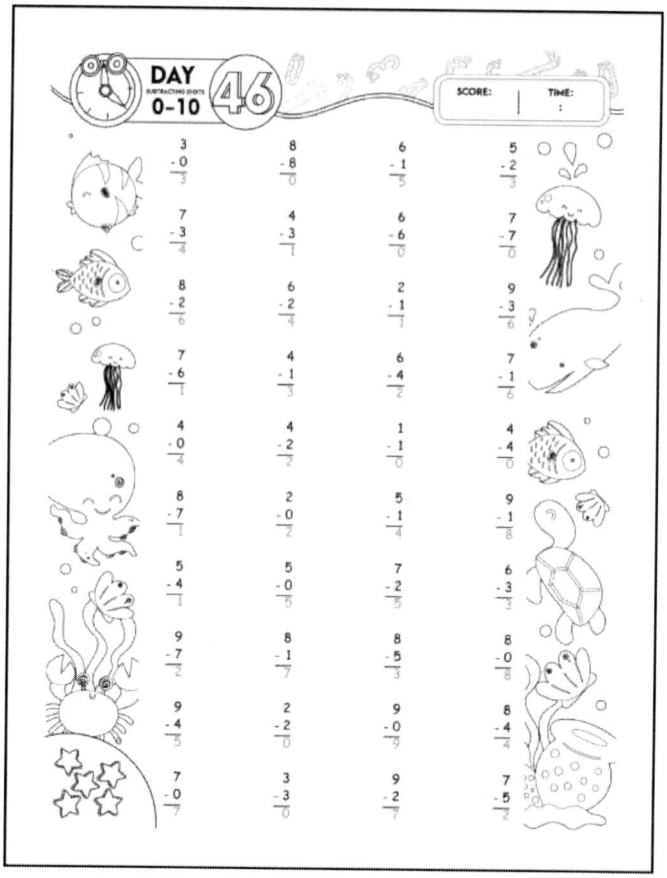

DAY 46 SUBTRACTING DIGITS 0–10

SCORE: | TIME: :

3 - 0 = 3	8 - 8 = 0	6 - 1 = 5	5 - 2 = 3
7 - 3 = 4	4 - 3 = 1	6 - 6 = 0	7 - 7 = 0
8 - 2 = 6	6 - 2 = 4	2 - 1 = 1	9 - 3 = 6
7 - 6 = 1	4 - 1 = 3	6 - 4 = 2	7 - 1 = 6
4 - 0 = 4	4 - 2 = 2	1 - 1 = 0	4 - 4 = 0
8 - 7 = 1	2 - 0 = 2	5 - 1 = 4	9 - 1 = 8
5 - 4 = 1	5 - 0 = 5	7 - 2 = 5	6 - 3 = 3
9 - 7 = 2	8 - 1 = 7	8 - 5 = 3	8 - 0 = 8
9 - 4 = 5	2 - 2 = 0	9 - 0 = 9	8 - 4 = 4
7 - 0 = 7	3 - 3 = 0	9 - 2 = 7	7 - 5 = 2

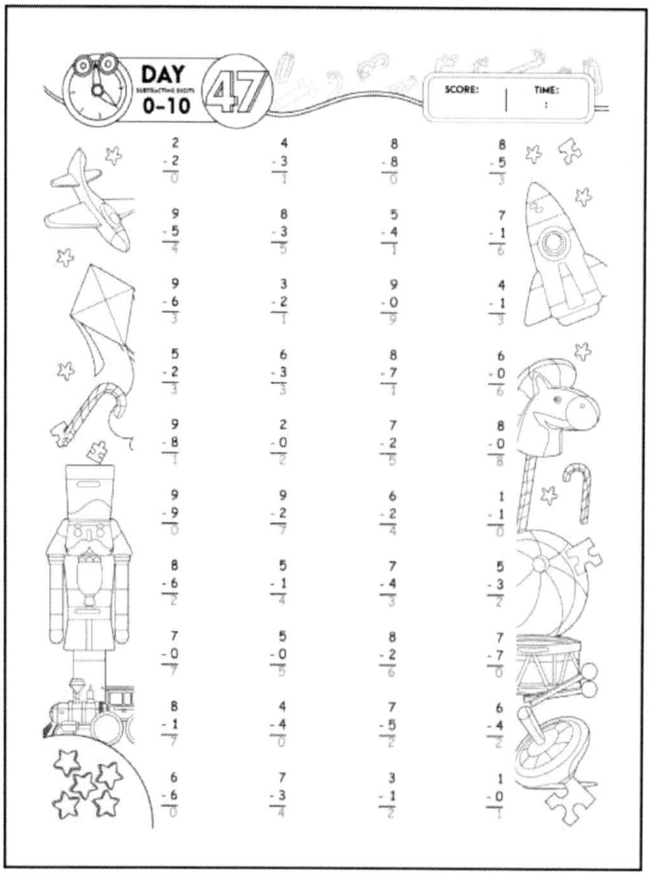

DAY 47 SUBTRACTING DIGITS 0–10

SCORE: | TIME: :

2 - 2 = 0	4 - 3 = 1	8 - 8 = 0	8 - 5 = 3
9 - 5 = 4	8 - 3 = 5	5 - 4 = 1	7 - 1 = 6
9 - 6 = 3	3 - 2 = 1	9 - 0 = 9	4 - 1 = 3
5 - 2 = 3	6 - 3 = 3	8 - 7 = 1	6 - 0 = 6
9 - 8 = 1	2 - 0 = 2	7 - 2 = 5	8 - 0 = 8
9 - 9 = 0	9 - 2 = 7	6 - 2 = 4	1 - 1 = 0
8 - 6 = 2	5 - 1 = 4	7 - 4 = 3	5 - 3 = 2
7 - 0 = 7	5 - 0 = 5	8 - 2 = 6	7 - 7 = 0
8 - 1 = 7	4 - 4 = 0	7 - 5 = 2	6 - 4 = 2
6 - 6 = 0	7 - 3 = 4	3 - 1 = 2	1 - 0 = 1

DAY 48 SUBTRACTING DIGITS 0–10

SCORE: | TIME: :

6 - 2 = 4	9 - 1 = 8	4 - 1 = 3	7 - 2 = 5
5 - 2 = 3	6 - 0 = 6	8 - 2 = 6	9 - 4 = 5
4 - 3 = 1	6 - 4 = 2	9 - 5 = 4	5 - 0 = 5
2 - 1 = 1	8 - 3 = 5	8 - 5 = 3	9 - 3 = 6
6 - 6 = 0	5 - 4 = 1	7 - 5 = 2	9 - 9 = 0
7 - 0 = 7	5 - 1 = 4	2 - 2 = 0	6 - 1 = 5
9 - 8 = 1	8 - 6 = 2	4 - 2 = 2	6 - 3 = 3
3 - 0 = 3	5 - 3 = 2	8 - 4 = 4	3 - 3 = 0
7 - 4 = 3	8 - 8 = 0	2 - 0 = 2	3 - 2 = 1
9 - 2 = 7	4 - 0 = 4	3 - 1 = 2	9 - 7 = 2

ANSWER DAY 49-52

DAY 49
SUBTRACTING DIGITS 0-15

SCORE: TIME: :

9 - 5 = 4	13 - 13 = 0	11 - 4 = 7	11 - 11 = 0
10 - 7 = 3	13 - 4 = 9	8 - 8 = 0	7 - 1 = 6
7 - 5 = 2	9 - 6 = 3	9 - 4 = 5	11 - 5 = 6
12 - 7 = 5	14 - 13 = 1	8 - 5 = 3	5 - 5 = 0
4 - 1 = 3	14 - 2 = 12	11 - 7 = 4	9 - 1 = 8
12 - 11 = 1	12 - 3 = 9	5 - 1 = 4	8 - 1 = 7
5 - 0 = 5	11 - 1 = 10	6 - 4 = 2	1 - 0 = 1
13 - 6 = 7	10 - 3 = 7	13 - 1 = 12	10 - 0 = 10
14 - 8 = 6	9 - 7 = 2	6 - 6 = 0	6 - 3 = 3
12 - 0 = 12	15 - 15 = 0	15 - 14 = 1	12 - 5 = 7

DAY 50
SUBTRACTING DIGITS 0-15

SCORE: TIME: :

12 - 12 = 0	14 - 1 = 13	14 - 10 = 4	13 - 1 = 12
11 - 6 = 5	2 - 2 = 0	10 - 0 = 10	5 - 4 = 1
14 - 8 = 6	13 - 3 = 10	3 - 0 = 3	14 - 9 = 5
9 - 5 = 4	13 - 12 = 1	7 - 6 = 1	11 - 7 = 4
10 - 4 = 6	14 - 3 = 11	10 - 3 = 7	8 - 0 = 8
10 - 7 = 3	11 - 5 = 6	4 - 1 = 3	12 - 10 = 2
6 - 1 = 5	3 - 2 = 1	6 - 4 = 2	13 - 10 = 3
15 - 2 = 13	15 - 13 = 2	9 - 8 = 1	12 - 4 = 8
10 - 1 = 9	10 - 2 = 8	13 - 0 = 13	14 - 4 = 10
14 - 7 = 7	14 - 0 = 14	12 - 3 = 9	7 - 1 = 6

DAY 51
SUBTRACTING DIGITS 0-15

SCORE: TIME: :

15 - 12 = 3	12 - 0 = 12	5 - 4 = 1	15 - 0 = 15
11 - 8 = 3	5 - 5 = 0	15 - 5 = 10	14 - 10 = 4
15 - 3 = 12	6 - 5 = 1	10 - 10 = 0	11 - 4 = 7
14 - 0 = 14	2 - 1 = 1	8 - 2 = 6	8 - 5 = 3
9 - 7 = 2	3 - 3 = 0	3 - 2 = 1	13 - 5 = 8
12 - 3 = 9	14 - 1 = 13	12 - 6 = 6	2 - 0 = 2
9 - 1 = 8	4 - 2 = 2	3 - 0 = 3	10 - 6 = 4
9 - 9 = 0	5 - 2 = 3	10 - 3 = 7	6 - 6 = 0
11 - 0 = 11	9 - 6 = 3	15 - 11 = 4	7 - 1 = 6
3 - 1 = 2	15 - 9 = 6	10 - 4 = 6	8 - 1 = 7

DAY 52
SUBTRACTING DIGITS 0-15

SCORE: TIME: :

15 - 3 = 12	7 - 7 = 0	11 - 1 = 10	14 - 8 = 6
15 - 0 = 15	13 - 3 = 10	7 - 3 = 4	12 - 7 = 5
8 - 4 = 4	12 - 5 = 7	14 - 10 = 4	10 - 7 = 3
14 - 13 = 1	15 - 1 = 14	2 - 2 = 0	14 - 12 = 2
14 - 3 = 11	11 - 9 = 2	14 - 4 = 10	13 - 4 = 9
15 - 8 = 7	12 - 4 = 8	12 - 10 = 2	13 - 2 = 11
12 - 2 = 10	15 - 5 = 10	11 - 11 = 0	11 - 5 = 6
11 - 4 = 7	10 - 8 = 2	11 - 0 = 11	10 - 6 = 4
13 - 9 = 4	14 - 6 = 8	14 - 14 = 0	13 - 12 = 1
1 - 0 = 1	13 - 11 = 2	5 - 1 = 4	15 - 6 = 9

ANSWER DAY 53-56

DAY 53 — SUBTRACTING DIGITS 0–15

13 - 3 = 10	10 - 8 = 2	5 - 2 = 3	15 - 4 = 11
10 - 5 = 5	14 - 10 = 4	12 - 1 = 11	15 - 10 = 5
6 - 3 = 3	10 - 9 = 1	8 - 5 = 3	13 - 4 = 9
2 - 0 = 2	6 - 1 = 5	10 - 3 = 7	10 - 4 = 6
15 - 2 = 13	9 - 2 = 7	12 - 9 = 3	10 - 1 = 9
9 - 7 = 2	14 - 13 = 1	12 - 12 = 0	4 - 1 = 3
12 - 7 = 5	13 - 13 = 0	6 - 2 = 4	8 - 2 = 6
11 - 11 = 0	11 - 1 = 10	14 - 9 = 5	8 - 1 = 7
13 - 0 = 13	7 - 7 = 0	7 - 1 = 6	12 - 0 = 12
4 - 4 = 0	10 - 10 = 0	4 - 2 = 2	6 - 5 = 1

DAY 54 — SUBTRACTING DIGITS 0–15

11 - 4 = 7	10 - 7 = 3	8 - 4 = 4	11 - 2 = 9
13 - 0 = 13	6 - 0 = 6	12 - 1 = 11	7 - 1 = 6
6 - 4 = 2	14 - 14 = 0	12 - 11 = 1	12 - 4 = 8
11 - 3 = 8	14 - 5 = 9	11 - 1 = 10	3 - 3 = 0
13 - 9 = 4	15 - 15 = 0	15 - 8 = 7	15 - 4 = 11
15 - 10 = 5	12 - 7 = 5	11 - 11 = 0	13 - 10 = 3
8 - 1 = 7	1 - 1 = 0	9 - 7 = 2	14 - 3 = 11
7 - 2 = 5	2 - 2 = 0	12 - 8 = 4	10 - 8 = 2
10 - 5 = 5	15 - 1 = 14	10 - 4 = 6	4 - 3 = 1
15 - 11 = 4	14 - 11 = 3	7 - 3 = 4	9 - 1 = 8

DAY 55 — SUBTRACTING DIGITS 0–15

2 - 1 = 1	10 - 0 = 10	14 - 4 = 10	14 - 10 = 4
12 - 2 = 10	12 - 9 = 3	14 - 0 = 14	11 - 3 = 8
13 - 1 = 12	13 - 9 = 4	13 - 0 = 13	7 - 3 = 4
13 - 12 = 1	3 - 0 = 3	15 - 12 = 3	6 - 6 = 0
15 - 9 = 6	5 - 1 = 4	15 - 11 = 4	1 - 0 = 1
14 - 14 = 0	12 - 0 = 12	3 - 2 = 1	13 - 2 = 11
13 - 6 = 7	14 - 6 = 8	11 - 11 = 0	9 - 0 = 9
9 - 4 = 5	12 - 7 = 5	10 - 9 = 1	4 - 4 = 0
10 - 2 = 8	8 - 8 = 0	2 - 0 = 2	7 - 1 = 6
13 - 4 = 9	12 - 6 = 6	14 - 13 = 1	5 - 3 = 2

DAY 56 — SUBTRACTING DIGITS 0–20

5 - 0 = 5	16 - 15 = 1	16 - 0 = 16	10 - 5 = 5
7 - 2 = 5	17 - 4 = 13	3 - 3 = 0	20 - 12 = 8
17 - 0 = 17	10 - 10 = 0	9 - 5 = 4	5 - 4 = 1
18 - 8 = 10	8 - 7 = 1	16 - 2 = 14	4 - 4 = 0
7 - 7 = 0	18 - 5 = 13	20 - 20 = 0	16 - 11 = 5
18 - 13 = 5	12 - 4 = 8	4 - 3 = 1	15 - 9 = 6
11 - 10 = 1	17 - 1 = 16	20 - 15 = 5	14 - 5 = 9
19 - 1 = 18	13 - 11 = 2	15 - 10 = 5	19 - 5 = 14
15 - 0 = 15	15 - 5 = 10	14 - 0 = 14	13 - 7 = 6
19 - 7 = 12	7 - 4 = 3	10 - 1 = 9	2 - 2 = 0

ANSWER DAY 57-60

DAY 57
SUBTRACTING DIGITS 0-20

SCORE: | TIME: :

11 - 1 = 10	16 - 14 = 2	4 - 4 = 0	19 - 11 = 8
14 - 11 = 3	12 - 11 = 1	17 - 6 = 11	12 - 5 = 7
18 - 14 = 4	2 - 2 = 0	20 - 16 = 4	18 - 17 = 1
7 - 7 = 0	16 - 11 = 5	14 - 1 = 13	20 - 18 = 2
16 - 8 = 8	12 - 12 = 0	17 - 14 = 3	15 - 11 = 4
9 - 0 = 9	17 - 17 = 0	12 - 7 = 5	16 - 10 = 6
16 - 0 = 16	17 - 10 = 7	19 - 14 = 5	15 - 1 = 14
18 - 3 = 15	12 - 3 = 9	3 - 3 = 0	2 - 0 = 2
19 - 18 = 1	8 - 6 = 2	11 - 0 = 11	7 - 4 = 3
13 - 12 = 1	10 - 2 = 8	14 - 12 = 2	20 - 4 = 16

DAY 58
SUBTRACTING DIGITS 0-20

SCORE: | TIME: :

18 - 7 = 11	20 - 17 = 3	19 - 9 = 10	8 - 0 = 8
20 - 3 = 17	16 - 4 = 12	5 - 2 = 3	20 - 18 = 2
19 - 0 = 19	20 - 6 = 14	15 - 9 = 6	12 - 2 = 10
11 - 0 = 11	8 - 2 = 6	15 - 5 = 10	19 - 15 = 4
16 - 15 = 1	17 - 9 = 8	14 - 3 = 11	13 - 4 = 9
17 - 17 = 0	13 - 10 = 3	3 - 2 = 1	18 - 16 = 2
14 - 13 = 1	10 - 0 = 10	19 - 19 = 0	7 - 1 = 6
20 - 10 = 10	17 - 5 = 12	18 - 9 = 9	14 - 14 = 0
20 - 7 = 13	6 - 6 = 0	16 - 12 = 4	15 - 4 = 11
16 - 9 = 7	19 - 14 = 5	13 - 9 = 4	11 - 1 = 10

DAY 59
SUBTRACTING DIGITS 0-20

SCORE: | TIME: :

15 - 11 = 4	6 - 5 = 1	8 - 3 = 5	17 - 3 = 14
5 - 2 = 3	10 - 8 = 2	19 - 19 = 0	17 - 1 = 16
20 - 8 = 12	20 - 6 = 14	14 - 2 = 12	10 - 0 = 10
15 - 13 = 2	2 - 1 = 1	10 - 1 = 9	8 - 6 = 2
13 - 2 = 11	20 - 3 = 17	4 - 4 = 0	20 - 5 = 15
20 - 9 = 11	7 - 4 = 3	17 - 14 = 3	14 - 7 = 7
18 - 10 = 8	20 - 10 = 10	14 - 13 = 1	19 - 4 = 15
13 - 13 = 0	19 - 6 = 13	13 - 0 = 13	13 - 9 = 4
13 - 3 = 10	15 - 8 = 7	8 - 8 = 0	18 - 4 = 14
16 - 12 = 4	11 - 10 = 1	5 - 4 = 1	18 - 15 = 3

DAY 60
SUBTRACTING DIGITS 0-20

SCORE: | TIME: :

16 - 16 = 0	15 - 4 = 11	12 - 2 = 10	13 - 8 = 5
7 - 0 = 7	20 - 2 = 18	17 - 0 = 17	16 - 7 = 9
16 - 9 = 7	8 - 6 = 2	16 - 1 = 15	3 - 3 = 0
18 - 1 = 17	7 - 6 = 1	15 - 5 = 10	20 - 9 = 11
10 - 2 = 8	11 - 4 = 7	14 - 8 = 6	16 - 8 = 8
20 - 6 = 14	16 - 0 = 16	20 - 12 = 8	3 - 2 = 1
16 - 3 = 13	10 - 4 = 6	8 - 4 = 4	17 - 3 = 14
5 - 0 = 5	16 - 15 = 1	14 - 14 = 0	6 - 1 = 5
13 - 9 = 4	7 - 2 = 5	14 - 5 = 9	16 - 2 = 14
17 - 8 = 9	20 - 4 = 16	19 - 16 = 3	9 - 5 = 4

ANSWER DAY 61-64

DAY 61

SUBTRACTING DIGITS 0-20

SCORE: | TIME: :

10 - 0 10	16 - 2 14	10 - 7 3	20 - 16 4
5 - 0 5	16 - 0 16	16 - 6 10	20 - 5 15
15 - 1 14	17 - 11 6	4 - 2 2	20 - 17 3
13 - 1 12	15 - 11 4	14 - 1 13	9 - 6 3
18 - 9 9	17 - 16 1	10 - 3 7	15 - 12 3
11 - 8 3	17 - 0 17	17 - 4 13	5 - 3 2
19 - 8 11	13 - 0 13	6 - 6 0	9 - 8 1
12 - 4 8	4 - 3 1	10 - 1 9	15 - 14 1
17 - 12 5	14 - 3 11	6 - 2 4	9 - 1 8
16 - 16 0	12 - 2 10	8 - 2 6	17 - 5 12

DAY 62

SUBTRACTING DIGITS 0-20

SCORE: | TIME: :

8 - 0 8	16 - 1 15	14 - 12 2	17 - 4 13
14 - 5 9	13 - 1 12	20 - 6 14	19 - 9 10
17 - 11 6	20 - 0 20	19 - 8 11	9 - 0 9
2 - 1 1	18 - 6 12	19 - 11 8	20 - 12 8
9 - 8 1	7 - 3 4	7 - 0 7	16 - 4 12
11 - 0 11	16 - 11 5	10 - 3 7	17 - 1 16
8 - 7 1	14 - 9 5	12 - 1 11	20 - 20 0
11 - 10 1	18 - 9 9	16 - 16 0	8 - 3 5
11 - 1 10	14 - 4 10	10 - 7 3	16 - 2 14
17 - 2 15	20 - 5 15	18 - 5 13	20 - 4 16

DAY 63

SUBTRACTING DIGITS 0-20

SCORE: | TIME: :

16 - 14 2	12 - 10 2	15 - 9 6	2 - 2 0
6 - 5 1	10 - 0 10	17 - 14 3	11 - 1 10
7 - 5 2	16 - 4 12	8 - 8 0	14 - 1 13
17 - 8 9	7 - 6 1	8 - 6 2	6 - 1 5
5 - 4 1	5 - 3 2	19 - 2 17	16 - 11 5
13 - 2 11	18 - 11 7	10 - 3 7	18 - 4 14
4 - 4 0	17 - 3 14	15 - 14 1	16 - 12 4
15 - 1 14	19 - 13 6	20 - 19 1	6 - 0 6
3 - 3 0	18 - 10 8	16 - 3 13	18 - 16 2
16 - 5 11	5 - 5 0	11 - 9 2	19 - 9 10

DAY 64

SUBTRACTING DIGITS 0-20

SCORE: | TIME: :

17 - 15 2	6 - 0 6	18 - 17 1	20 - 4 16
9 - 6 3	20 - 12 8	9 - 9 0	14 - 5 9
18 - 2 16	19 - 9 10	8 - 5 3	14 - 0 14
7 - 5 2	18 - 18 0	19 - 7 12	16 - 11 5
19 - 17 2	17 - 7 10	19 - 5 14	20 - 20 0
7 - 4 3	13 - 7 6	11 - 2 9	17 - 0 17
19 - 16 3	13 - 2 11	10 - 9 1	14 - 12 2
14 - 1 13	20 - 10 10	9 - 4 5	18 - 4 14
17 - 9 8	16 - 8 8	8 - 4 4	13 - 13 0
7 - 2 5	19 - 15 4	15 - 7 8	15 - 15 0

ANSWER DAY 65-68

DAY 65 — SUBTRACTING DIGITS 0-20

SCORE: | TIME: :

17 - 6 = 11	6 - 1 = 5	19 - 10 = 9	19 - 0 = 19
20 - 12 = 8	17 - 5 = 12	10 - 0 = 10	15 - 6 = 9
19 - 8 = 11	9 - 4 = 5	18 - 5 = 13	8 - 2 = 6
11 - 3 = 8	17 - 2 = 15	14 - 6 = 8	15 - 0 = 15
10 - 9 = 1	18 - 12 = 6	5 - 5 = 0	9 - 9 = 0
12 - 9 = 3	2 - 2 = 0	20 - 0 = 20	15 - 10 = 5
7 - 5 = 2	19 - 3 = 16	19 - 9 = 10	9 - 0 = 9
17 - 3 = 14	10 - 6 = 4	4 - 3 = 1	2 - 0 = 2
3 - 1 = 2	17 - 14 = 3	6 - 4 = 2	12 - 10 = 2
20 - 15 = 5	19 - 6 = 13	15 - 14 = 1	13 - 5 = 8

DAY 66 — SUBTRACTING DIGITS 0-20

SCORE: | TIME: :

15 - 3 = 12	13 - 1 = 12	3 - 3 = 0	11 - 3 = 8
13 - 3 = 10	18 - 10 = 8	14 - 11 = 3	18 - 16 = 2
7 - 4 = 3	15 - 7 = 8	19 - 16 = 3	14 - 13 = 1
2 - 2 = 0	17 - 0 = 17	7 - 2 = 5	20 - 14 = 6
13 - 7 = 6	8 - 1 = 7	17 - 17 = 0	18 - 9 = 9
13 - 11 = 2	15 - 6 = 9	8 - 6 = 2	15 - 14 = 1
7 - 5 = 2	12 - 8 = 4	8 - 7 = 1	13 - 10 = 3
19 - 9 = 10	6 - 2 = 4	10 - 3 = 7	19 - 1 = 18
14 - 4 = 10	5 - 3 = 2	18 - 17 = 1	18 - 7 = 11
10 - 2 = 8	19 - 17 = 2	13 - 12 = 1	15 - 1 = 14

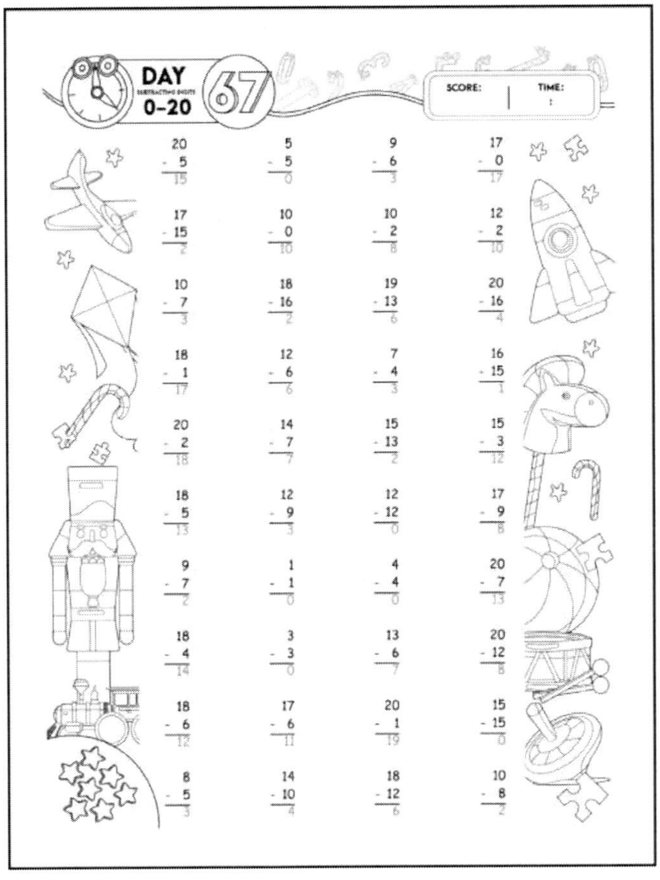

DAY 67 — SUBTRACTING DIGITS 0-20

SCORE: | TIME: :

20 - 5 = 15	5 - 5 = 0	9 - 6 = 3	17 - 0 = 17
17 - 15 = 2	10 - 0 = 10	10 - 2 = 8	12 - 2 = 10
10 - 7 = 3	18 - 16 = 2	19 - 13 = 6	20 - 16 = 4
18 - 1 = 17	12 - 6 = 6	7 - 4 = 3	16 - 15 = 1
20 - 2 = 18	14 - 7 = 7	15 - 13 = 2	15 - 3 = 12
18 - 5 = 13	12 - 9 = 3	12 - 12 = 0	17 - 9 = 8
9 - 7 = 2	1 - 1 = 0	4 - 4 = 0	20 - 7 = 13
18 - 4 = 14	3 - 3 = 0	13 - 6 = 7	20 - 12 = 8
18 - 6 = 12	17 - 6 = 11	20 - 1 = 19	15 - 15 = 0
8 - 5 = 3	14 - 10 = 4	18 - 12 = 6	10 - 8 = 2

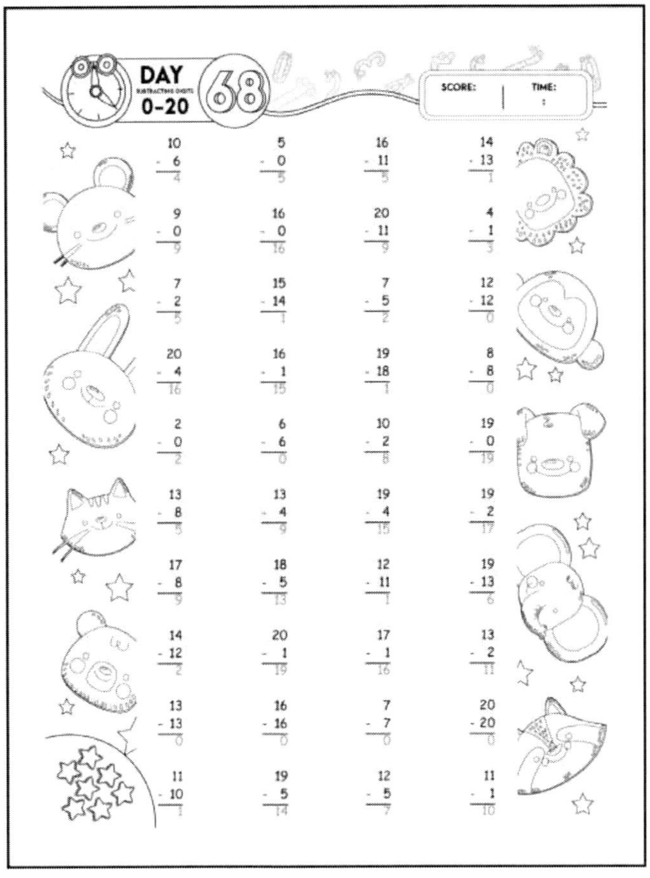

DAY 68 — SUBTRACTING DIGITS 0-20

SCORE: | TIME: :

10 - 6 = 4	5 - 0 = 5	16 - 11 = 5	14 - 13 = 1
9 - 0 = 9	16 - 0 = 16	20 - 11 = 9	4 - 1 = 3
7 - 2 = 5	15 - 14 = 1	7 - 5 = 2	12 - 12 = 0
20 - 4 = 16	16 - 1 = 15	19 - 18 = 1	8 - 8 = 0
2 - 0 = 2	6 - 6 = 0	10 - 2 = 8	19 - 0 = 19
13 - 8 = 5	13 - 4 = 9	19 - 4 = 15	19 - 2 = 17
17 - 8 = 9	18 - 5 = 13	12 - 11 = 1	19 - 13 = 6
14 - 12 = 2	20 - 1 = 19	17 - 1 = 16	13 - 2 = 11
13 - 13 = 0	16 - 16 = 0	7 - 7 = 0	20 - 20 = 0
11 - 10 = 1	19 - 5 = 14	12 - 5 = 7	11 - 1 = 10

ANSWER DAY 69-72

DAY 69
SUBTRACTING DIGITS 0-20

SCORE: | TIME: :

19 - 14 = 5	11 - 10 = 1	7 - 3 = 4	14 - 11 = 3
18 - 16 = 2	17 - 4 = 13	9 - 7 = 2	19 - 6 = 13
14 - 14 = 0	14 - 3 = 11	14 - 2 = 12	13 - 2 = 11
18 - 5 = 13	10 - 0 = 10	8 - 1 = 7	17 - 16 = 1
13 - 10 = 3	14 - 9 = 5	13 - 6 = 7	13 - 4 = 9
20 - 10 = 10	18 - 11 = 7	17 - 1 = 16	12 - 12 = 0
19 - 7 = 12	20 - 17 = 3	20 - 7 = 13	15 - 9 = 6
16 - 8 = 8	17 - 10 = 7	17 - 13 = 4	17 - 12 = 5
20 - 5 = 15	13 - 5 = 8	14 - 10 = 4	10 - 5 = 5
12 - 8 = 4	20 - 2 = 18	14 - 13 = 1	15 - 5 = 10

DAY 70
SUBTRACTING DIGITS 0-20

SCORE: | TIME: :

16 - 9 = 7	11 - 4 = 7	16 - 4 = 12	19 - 15 = 4
14 - 8 = 6	13 - 2 = 11	20 - 4 = 16	10 - 2 = 8
19 - 16 = 3	10 - 5 = 5	19 - 19 = 0	16 - 2 = 14
13 - 6 = 7	20 - 5 = 15	17 - 14 = 3	4 - 1 = 3
2 - 2 = 0	16 - 12 = 4	18 - 0 = 18	2 - 0 = 2
7 - 6 = 1	6 - 4 = 2	8 - 8 = 0	16 - 15 = 1
19 - 1 = 18	9 - 9 = 0	6 - 3 = 3	4 - 2 = 2
17 - 7 = 10	10 - 9 = 1	14 - 11 = 3	17 - 5 = 12
16 - 8 = 8	18 - 10 = 8	10 - 0 = 10	16 - 7 = 9
17 - 16 = 1	15 - 8 = 7	11 - 9 = 2	13 - 12 = 1

DAY 71
SUBTRACTING DIGITS 0-20

SCORE: | TIME: :

9 - 6 = 3	4 - 0 = 4	13 - 4 = 9	14 - 3 = 11
16 - 3 = 13	15 - 4 = 11	5 - 0 = 5	18 - 6 = 12
18 - 7 = 11	17 - 11 = 6	10 - 10 = 0	15 - 8 = 7
15 - 7 = 8	17 - 10 = 7	14 - 13 = 1	6 - 5 = 1
13 - 1 = 12	3 - 1 = 2	8 - 1 = 7	12 - 10 = 2
20 - 19 = 1	14 - 0 = 14	14 - 14 = 0	13 - 12 = 1
12 - 11 = 1	12 - 4 = 8	8 - 0 = 8	20 - 16 = 4
7 - 2 = 5	14 - 1 = 13	19 - 19 = 0	2 - 0 = 2
20 - 14 = 6	10 - 4 = 6	19 - 7 = 12	16 - 0 = 16
12 - 8 = 4	8 - 6 = 2	18 - 16 = 2	15 - 10 = 5

DAY 72
SUBTRACTING DIGITS 0-20

SCORE: | TIME: :

12 - 9 = 3	7 - 6 = 1	14 - 9 = 5	16 - 0 = 16
20 - 16 = 4	16 - 6 = 10	4 - 3 = 1	5 - 5 = 0
8 - 1 = 7	14 - 14 = 0	3 - 2 = 1	18 - 14 = 4
14 - 7 = 7	9 - 2 = 7	7 - 5 = 2	10 - 10 = 0
3 - 3 = 0	20 - 5 = 15	17 - 0 = 17	16 - 9 = 7
20 - 20 = 0	12 - 1 = 11	12 - 12 = 0	14 - 2 = 12
19 - 19 = 0	11 - 10 = 1	4 - 4 = 0	7 - 2 = 5
7 - 4 = 3	20 - 1 = 19	18 - 16 = 2	9 - 7 = 2
8 - 7 = 1	10 - 1 = 9	13 - 7 = 6	6 - 0 = 6
12 - 7 = 5	10 - 8 = 2	19 - 1 = 18	9 - 6 = 3

ANSWER DAY 73-76

DAY 73 SUBTRACTING DIGITS 0-20 — SCORE: | TIME: :

16 - 4 = 12	8 - 0 = 8	19 - 14 = 5	15 - 5 = 10
11 - 9 = 2	18 - 8 = 10	8 - 7 = 1	14 - 1 = 13
13 - 11 = 2	4 - 1 = 3	5 - 1 = 4	15 - 13 = 2
16 - 13 = 3	13 - 4 = 9	18 - 9 = 9	10 - 2 = 8
6 - 5 = 1	12 - 12 = 0	19 - 5 = 14	20 - 17 = 3
4 - 4 = 0	20 - 3 = 17	14 - 8 = 6	8 - 6 = 2
4 - 0 = 4	20 - 11 = 9	20 - 1 = 19	20 - 2 = 18
18 - 18 = 0	17 - 13 = 4	12 - 0 = 12	9 - 9 = 0
14 - 0 = 14	19 - 10 = 9	9 - 7 = 2	10 - 4 = 6
9 - 5 = 4	19 - 18 = 1	12 - 10 = 2	17 - 3 = 14

DAY 74 SUBTRACTING DIGITS 0-20 — SCORE: | TIME: :

16 - 11 = 5	17 - 4 = 13	20 - 6 = 14	20 - 18 = 2
12 - 12 = 0	11 - 9 = 2	16 - 9 = 7	9 - 7 = 2
1 - 0 = 1	19 - 17 = 2	10 - 8 = 2	9 - 1 = 8
19 - 14 = 5	20 - 11 = 9	16 - 5 = 11	19 - 6 = 13
18 - 15 = 3	20 - 7 = 13	17 - 2 = 15	9 - 6 = 3
11 - 1 = 10	16 - 13 = 3	18 - 17 = 1	20 - 8 = 12
14 - 4 = 10	6 - 2 = 4	20 - 14 = 6	17 - 8 = 9
10 - 4 = 6	7 - 7 = 0	14 - 9 = 5	13 - 6 = 7
19 - 3 = 16	11 - 8 = 3	9 - 4 = 5	10 - 7 = 3
7 - 4 = 3	5 - 1 = 4	7 - 1 = 6	15 - 12 = 3

DAY 75 SUBTRACTING DIGITS 0-20 — SCORE: | TIME: :

10 - 3 = 7	17 - 4 = 13	19 - 13 = 6	19 - 12 = 7
6 - 5 = 1	19 - 1 = 18	13 - 2 = 11	19 - 7 = 12
18 - 16 = 2	10 - 2 = 8	11 - 8 = 3	15 - 1 = 14
6 - 3 = 3	7 - 0 = 7	9 - 9 = 0	20 - 17 = 3
10 - 7 = 3	9 - 6 = 3	12 - 4 = 8	7 - 3 = 4
9 - 4 = 5	4 - 4 = 0	15 - 5 = 10	11 - 6 = 5
13 - 0 = 13	17 - 0 = 17	9 - 2 = 7	20 - 18 = 2
8 - 3 = 5	11 - 7 = 4	16 - 8 = 8	4 - 2 = 2
16 - 14 = 2	6 - 6 = 0	11 - 10 = 1	12 - 1 = 11
17 - 10 = 7	19 - 8 = 11	6 - 0 = 6	12 - 3 = 9

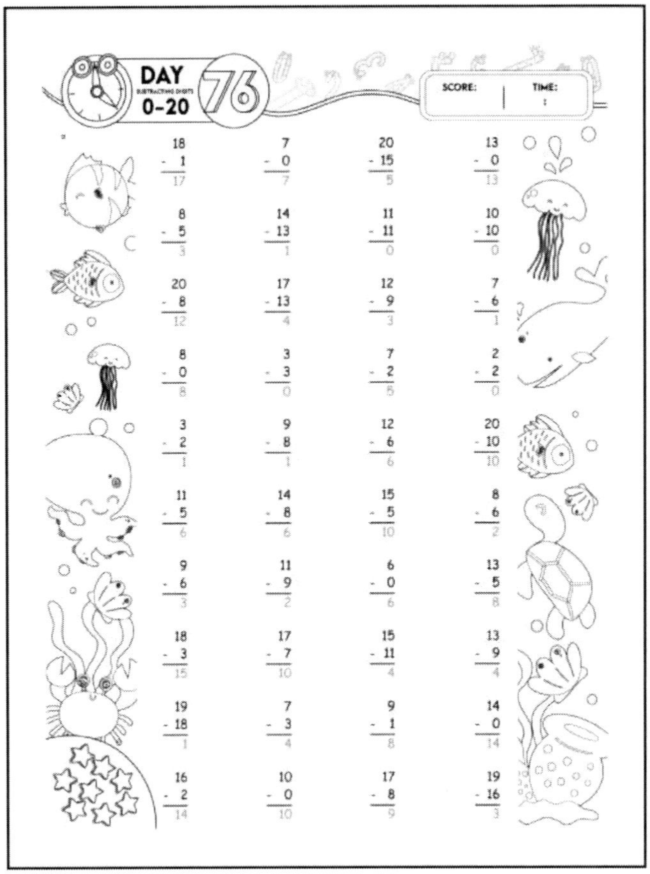

DAY 76 SUBTRACTING DIGITS 0-20 — SCORE: | TIME: :

18 - 1 = 17	7 - 0 = 7	20 - 15 = 5	13 - 0 = 13
8 - 5 = 3	14 - 13 = 1	11 - 11 = 0	10 - 10 = 0
20 - 8 = 12	17 - 13 = 4	12 - 9 = 3	7 - 6 = 1
8 - 0 = 8	3 - 3 = 0	7 - 2 = 5	2 - 2 = 0
3 - 2 = 1	9 - 8 = 1	12 - 6 = 6	20 - 10 = 10
11 - 5 = 6	14 - 8 = 6	15 - 5 = 10	8 - 6 = 2
9 - 6 = 3	11 - 9 = 2	6 - 0 = 6	13 - 5 = 8
18 - 3 = 15	17 - 7 = 10	15 - 11 = 4	13 - 9 = 4
19 - 18 = 1	7 - 3 = 4	9 - 1 = 8	14 - 0 = 14
16 - 2 = 14	10 - 0 = 10	17 - 8 = 9	19 - 16 = 3

ANSWER DAY 77-80

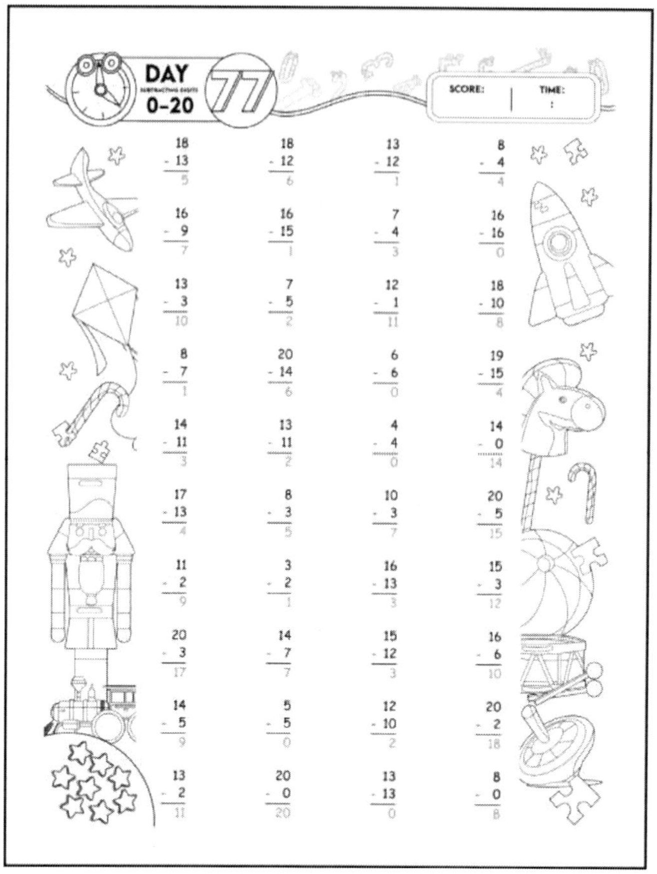

DAY 77 — SUBTRACTING DIGITS 0-20

SCORE: | TIME: :

18 − 13 = 5	18 − 12 = 6	13 − 12 = 1	8 − 4 = 4
16 − 9 = 7	16 − 15 = 1	7 − 4 = 3	16 − 16 = 0
13 − 3 = 10	7 − 5 = 2	12 − 1 = 11	18 − 10 = 8
8 − 7 = 1	20 − 14 = 6	6 − 6 = 0	19 − 15 = 4
14 − 11 = 3	13 − 11 = 2	4 − 4 = 0	14 − 0 = 14
17 − 13 = 4	8 − 3 = 5	10 − 3 = 7	20 − 5 = 15
11 − 2 = 9	3 − 2 = 1	16 − 13 = 3	15 − 3 = 12
20 − 3 = 17	14 − 7 = 7	15 − 12 = 3	16 − 6 = 10
14 − 5 = 9	5 − 5 = 0	12 − 10 = 2	20 − 2 = 18
13 − 2 = 11	20 − 0 = 20	13 − 13 = 0	8 − 0 = 8

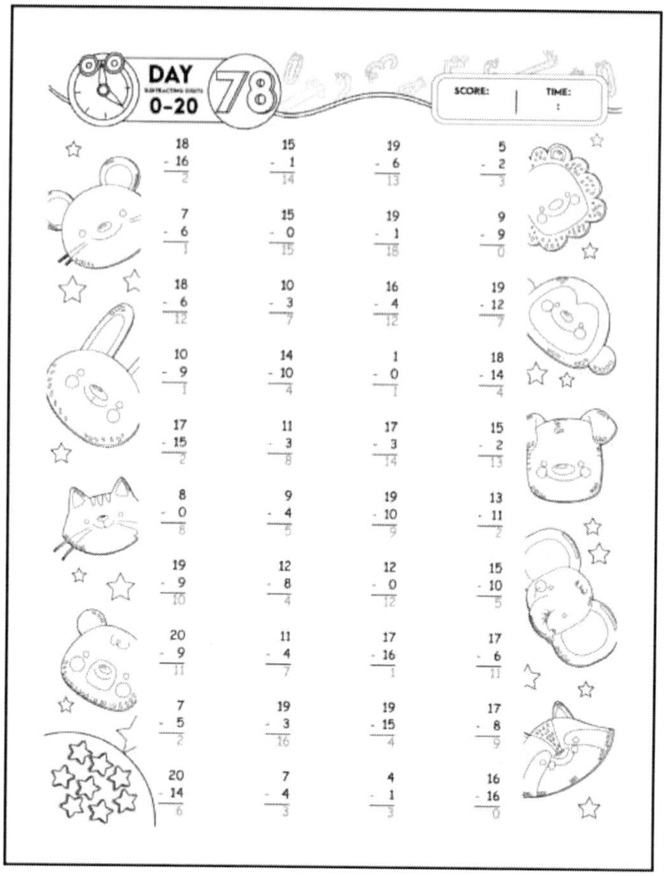

DAY 78 — SUBTRACTING DIGITS 0-20

SCORE: | TIME: :

18 − 16 = 2	15 − 1 = 14	19 − 6 = 13	5 − 2 = 3
7 − 6 = 1	15 − 0 = 15	19 − 1 = 18	9 − 9 = 0
18 − 6 = 12	10 − 3 = 7	16 − 4 = 12	19 − 12 = 7
10 − 9 = 1	14 − 10 = 4	1 − 0 = 1	18 − 14 = 4
17 − 15 = 2	11 − 3 = 8	17 − 3 = 14	15 − 2 = 13
8 − 0 = 8	9 − 4 = 5	19 − 10 = 9	13 − 11 = 2
19 − 9 = 10	12 − 8 = 4	12 − 0 = 12	15 − 10 = 5
20 − 9 = 11	11 − 4 = 7	17 − 16 = 1	17 − 6 = 11
7 − 5 = 2	19 − 3 = 16	19 − 15 = 4	17 − 8 = 9
20 − 14 = 6	7 − 4 = 3	4 − 1 = 3	16 − 16 = 0

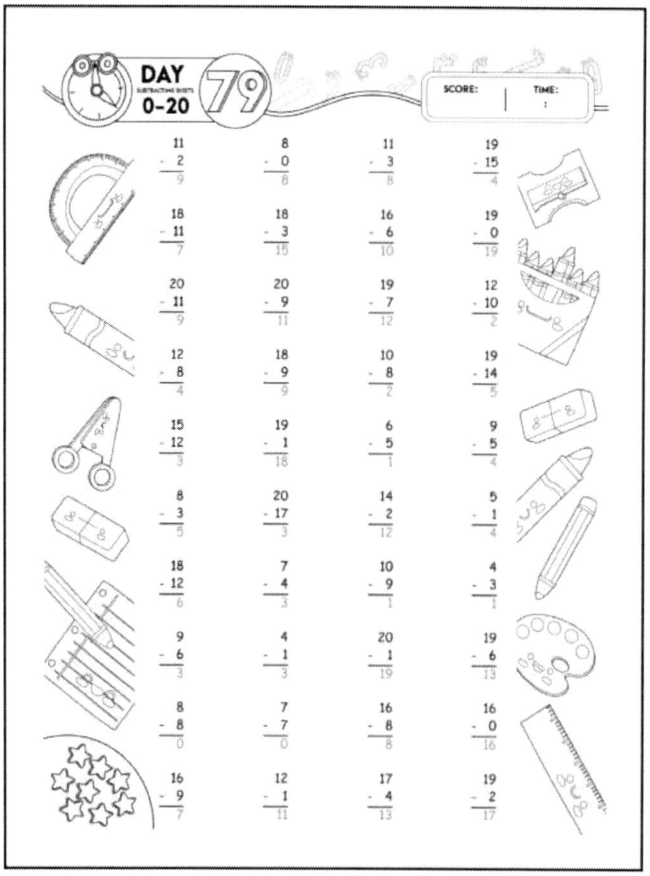

DAY 79 — SUBTRACTING DIGITS 0-20

SCORE: | TIME: :

11 − 2 = 9	8 − 0 = 8	11 − 3 = 8	19 − 15 = 4
18 − 11 = 7	18 − 3 = 15	16 − 6 = 10	19 − 0 = 19
20 − 11 = 9	20 − 9 = 11	19 − 7 = 12	12 − 10 = 2
12 − 8 = 4	18 − 9 = 9	10 − 8 = 2	19 − 14 = 5
15 − 12 = 3	19 − 1 = 18	6 − 5 = 1	9 − 5 = 4
8 − 3 = 5	20 − 17 = 3	14 − 2 = 12	5 − 1 = 4
18 − 12 = 6	7 − 4 = 3	10 − 9 = 1	4 − 3 = 1
9 − 6 = 3	4 − 1 = 3	20 − 1 = 19	19 − 6 = 13
8 − 8 = 0	7 − 7 = 0	16 − 8 = 8	16 − 0 = 16
16 − 9 = 7	12 − 1 = 11	17 − 4 = 13	19 − 2 = 17

DAY 80 — SUBTRACTING DIGITS 0-20

SCORE: | TIME: :

20 − 20 = 0	6 − 2 = 4	13 − 2 = 11	5 − 4 = 1
14 − 2 = 12	18 − 6 = 12	15 − 13 = 2	17 − 11 = 6
8 − 8 = 0	15 − 9 = 6	18 − 14 = 4	13 − 8 = 5
11 − 0 = 11	10 − 2 = 8	18 − 5 = 13	7 − 5 = 2
4 − 2 = 2	13 − 5 = 8	19 − 10 = 9	11 − 3 = 8
9 − 8 = 1	9 − 6 = 3	20 − 18 = 2	15 − 12 = 3
7 − 2 = 5	5 − 0 = 5	13 − 7 = 6	17 − 14 = 3
11 − 5 = 6	12 − 7 = 5	11 − 8 = 3	17 − 12 = 5
9 − 7 = 2	15 − 2 = 13	1 − 1 = 0	12 − 12 = 0
10 − 8 = 2	19 − 8 = 11	14 − 0 = 14	19 − 7 = 12

ANSWER DAY 81-84

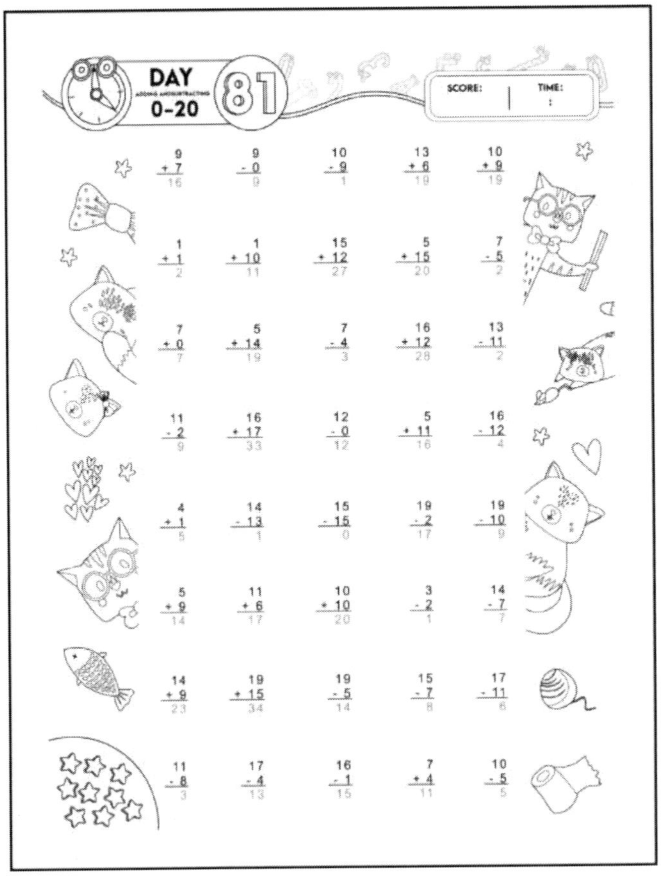
DAY 81
ADDING AND SUBTRACTING 0-20
SCORE: | TIME: :

9 + 7 16	9 - 0 9	10 - 9 1	13 + 6 19	10 + 9 19
1 + 1 2	1 + 10 11	15 + 12 27	5 + 15 20	7 - 5 2
7 + 0 7	5 + 14 19	7 - 4 3	16 + 12 28	13 - 11 2
11 - 2 9	16 + 17 33	12 - 0 12	5 + 11 16	16 - 12 4
4 + 1 5	14 - 13 1	15 - 15 0	19 - 2 17	19 - 10 9
5 + 9 14	11 + 6 17	10 + 10 20	3 - 2 1	14 - 7 7
14 + 9 23	19 + 15 34	19 - 5 14	15 - 7 8	17 - 11 6
11 - 8 3	17 - 4 13	16 - 1 15	7 + 4 11	10 - 5 5

DAY 82
ADDING AND SUBTRACTING 0-20
SCORE: | TIME: :

14 + 7 21	4 + 11 15	14 - 13 1	9 + 8 17	11 - 1 10
3 + 6 9	6 + 4 10	4 + 6 10	15 + 3 18	9 + 1 10
4 - 0 4	18 + 11 29	3 + 5 8	11 - 9 2	5 - 5 0
10 + 16 26	0 + 7 7	15 + 2 17	9 - 6 3	3 + 6 9
15 - 1 14	13 + 16 29	9 - 3 6	0 + 16 16	6 - 4 2
13 + 7 20	2 + 7 9	10 + 12 22	1 + 16 17	6 - 5 1
3 - 2 1	2 - 0 2	10 - 6 4	13 + 16 29	11 - 11 0
16 - 4 12	16 + 18 34	18 - 5 13	9 - 2 7	17 + 17 34

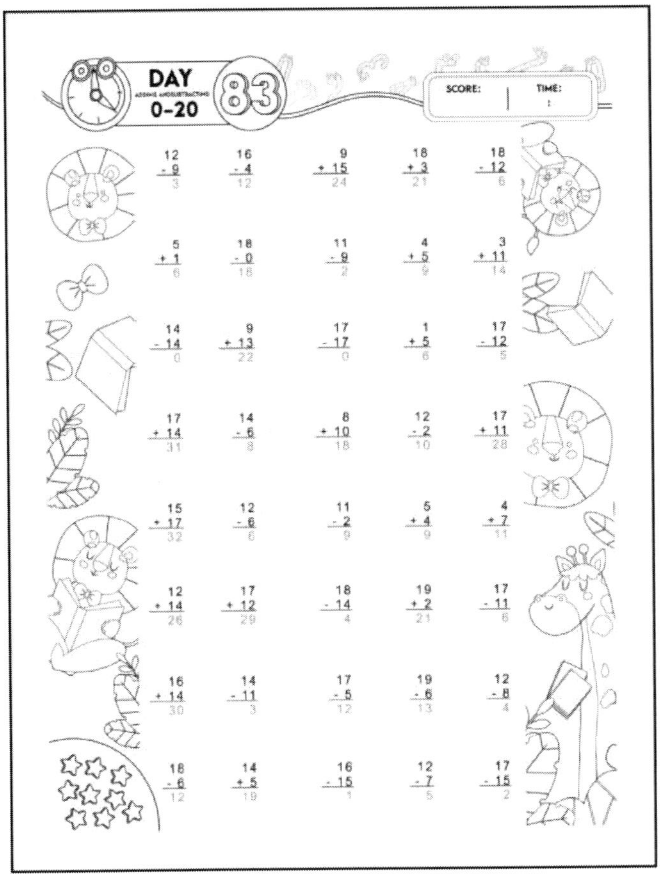
DAY 83
ADDING AND SUBTRACTING 0-20
SCORE: | TIME: :

12 - 9 3	16 - 4 12	9 + 15 24	18 + 3 21	18 - 12 6
5 + 1 6	18 - 0 18	11 - 9 2	4 + 5 9	3 + 11 14
14 - 14 0	9 + 13 22	17 - 17 0	1 + 5 6	17 - 12 5
17 + 14 31	14 - 6 8	8 + 10 18	12 - 2 10	17 + 11 28
15 + 17 32	12 - 6 6	11 - 2 9	5 + 4 9	4 + 7 11
12 + 14 26	17 + 12 29	18 - 14 4	19 + 2 21	17 - 11 6
16 + 14 30	14 - 11 3	17 - 5 12	19 - 6 13	12 - 8 4
18 - 6 12	14 + 5 19	16 - 15 1	12 - 7 5	17 - 15 2

DAY 84
ADDING AND SUBTRACTING 0-20
SCORE: | TIME: :

16 + 2 18	10 + 15 25	19 - 6 13	3 15 18	+ 15 - 7 8
7 + 10 17	13 - 13 0	1 + 16 17	8 + 4 12	18 + 5 23
16 - 3 13	17 - 7 10	18 - 0 18	18 17 35	+ 18 15 3
14 + 17 31	13 + 18 31	2 + 4 6	8 - 1 7	4 13 17
9 - 6 3	0 + 13 13	5 + 0 5	7 + 4 11	11 - 3 8
18 - 4 14	9 - 7 2	8 + 13 21	13 - 6 7	2 + 9 11
1 + 13 14	3 - 2 1	14 + 6 20	7 + 4 11	18 - 4 14
14 + 8 22	15 + 2 17	16 + 12 28	10 - 8 2	11 - 6 5

ANSWER DAY 85-89

DAY 85
ADDING AND SUBTRACTING 0-20

SCORE: | TIME: :

9 - 9 = 0	15 - 1 = 14	9 + 12 = 21	6 + 12 = 18	13 - 3 = 10
15 - 3 = 12	16 + 16 = 32	16 - 16 = 0	12 - 4 = 8	6 + 8 = 14
18 + 12 = 30	8 - 2 = 6	6 + 6 = 12	8 - 6 = 2	16 - 7 = 9
3 + 7 = 10	12 - 4 = 8	7 + 14 = 21	18 - 15 = 3	4 - 3 = 1
17 + 18 = 35	9 - 4 = 5	11 + 18 = 29	10 + 12 = 22	10 - 9 = 1
3 + 3 = 6	17 - 6 = 11	5 + 13 = 18	15 + 8 = 23	13 + 6 = 19
9 - 5 = 4	7 - 3 = 4	14 + 13 = 27	11 + 10 = 21	13 - 3 = 10
5 - 2 = 3	1 + 18 = 19	14 + 11 = 25	10 + 15 = 25	9 + 0 = 9

DAY 86
ADDING AND SUBTRACTING 0-20

SCORE: | TIME: :

16 + 12 = 28	18 + 1 = 19	16 + 11 = 27	3 + 4 = 7	18 + 13 = 31
2 + 12 = 14	19 - 11 = 8	9 - 3 = 6	9 + 4 = 13	17 - 5 = 12
3 + 16 = 19	7 - 1 = 6	9 - 1 = 8	11 - 5 = 6	17 - 11 = 6
15 + 8 = 23	3 + 2 = 5	14 + 18 = 32	13 + 17 = 30	13 - 2 = 11
18 - 18 = 0	9 + 10 = 19	1 + 0 = 1	19 - 8 = 11	8 - 4 = 4
12 - 4 = 8	6 - 5 = 1	5 - 3 = 2	18 + 7 = 25	7 - 3 = 4
7 - 2 = 5	17 - 3 = 14	17 - 8 = 9	12 - 11 = 1	5 + 10 = 15
0 + 2 = 2	15 - 7 = 8	18 - 8 = 10	15 + 0 = 15	18 - 17 = 1

DAY 87
ADDING AND SUBTRACTING 0-20

SCORE: | TIME: :

14 - 8 = 6	15 + 15 = 30	10 + 9 = 19	6 - 2 = 4	11 - 8 = 3
10 - 6 = 4	18 - 17 = 1	12 - 10 = 2	8 - 3 = 5	10 + 6 = 16
16 - 7 = 9	6 + 7 = 13	11 - 10 = 1	10 + 13 = 23	3 + 4 = 7
13 + 5 = 18	12 - 7 = 5	15 - 12 = 3	16 + 18 = 34	8 - 3 = 5
12 - 9 = 3	17 - 5 = 12	15 + 11 = 26	1 + 4 = 5	19 - 4 = 15
14 + 16 = 30	16 + 3 = 19	11 - 7 = 4	6 - 1 = 5	3 + 6 = 9
16 + 10 = 26	9 + 11 = 20	17 - 14 = 3	16 - 8 = 8	7 - 4 = 3
14 - 11 = 3	14 - 6 = 8	9 + 10 = 19	17 + 15 = 32	16 + 5 = 21

DAY 88
ADDING AND SUBTRACTING 0-20

SCORE: | TIME: :

20 + 1 = 21	10 + 13 = 23	10 - 4 = 6	20 - 1 = 19	10 - 6 = 4
10 + 1 = 11	20 - 6 = 14	10 + 18 = 28	20 - 7 = 13	10 + 12 = 22
10 - 0 = 10	10 - 6 = 4	10 + 10 = 20	20 + 1 = 21	10 - 3 = 7
10 - 3 = 7	10 + 4 = 14	10 - 8 = 2	10 + 8 = 18	10 - 3 = 7
10 - 4 = 6	20 - 12 = 8	10 + 2 = 12	20 - 2 = 18	10 - 3 = 7
20 - 13 = 7	10 + 2 = 12	10 - 13 = -3	10 - 4 = 6	20 + 3 = 23
10 - 1 = 9	20 + 17 = 37	10 - 1 = 9	10 - 2 = 8	10 - 8 = 2
20 - 5 = 15	10 + 15 = 25	10 + 16 = 26	10 - 2 = 8	10 - 12 = -2

ANSWER DAY 90-92

DAY 89 — ADDING AND SUBTRACTING 0-20

SCORE: | TIME: :

10 + 15 = 25	20 - 15 = 5	10 - 1 = 9	10 - 2 = 8	10 + 4 = 14
10 - 3 = 7	10 + 10 = 20	10 - 9 = 1	10 + 10 = 20	20 + 1 = 21
10 + 6 = 16	10 - 10 = 0	10 + 1 = 11	20 + 3 = 23	20 + 5 = 25
10 + 19 = 29	10 + 3 = 13	20 - 7 = 13	10 + 10 = 20	20 + 4 = 24
10 - 3 = 7	10 + 12 = 22	10 - 1 = 9	10 + 7 = 17	10 + 9 = 19
10 - 4 = 6	10 + 19 = 29	10 + 1 = 11	10 + 4 = 14	10 - 5 = 5
10 + 15 = 25	20 + 11 = 31	10 + 8 = 18	20 - 14 = 6	10 - 6 = 4
20 - 4 = 16	10 - 7 = 3	20 - 13 = 7	20 - 11 = 9	10 - 2 = 8

DAY 90 — ADDING AND SUBTRACTING 0-20

SCORE: | TIME: :

20 + 3 = 23	20 - 7 = 13	10 + 4 = 14	20 - 16 = 4	10 - 2 = 8
10 + 7 = 17	20 + 16 = 36	10 - 4 = 6	10 + 15 = 25	20 - 1 = 19
10 + 13 = 23	10 + 19 = 29	10 + 11 = 21	10 + 5 = 15	10 + 10 = 20
10 - 9 = 1	10 + 8 = 18	20 - 1 = 19	10 + 2 = 12	10 - 10 = 0
20 - 6 = 14	10 + 8 = 18	10 + 10 = 20	10 - 8 = 2	20 - 8 = 12
10 - 0 = 10	20 - 1 = 19	10 - 1 = 9	20 + 14 = 34	10 - 0 = 10
10 - 4 = 6	20 - 2 = 18	10 + 15 = 25	10 - 2 = 8	20 - 7 = 13
10 - 5 = 5	20 - 11 = 9	10 + 5 = 15	20 - 15 = 5	20 + 8 = 28

DAY 91 — ADDING AND SUBTRACTING 0-20

SCORE: | TIME: :

10 - 5 = 5	10 - 4 = 6	10 - 4 = 6	20 - 2 = 18	20 + 18 = 38
10 + 12 = 22	20 + 17 = 37	10 + 5 = 15	10 - 2 = 8	10 + 1 = 11
20 - 1 = 19	10 + 9 = 19	10 - 8 = 2	10 + 11 = 21	10 + 8 = 18
10 - 3 = 7	10 - 11 = -1	10 - 5 = 5	10 - 1 = 9	10 + 10 = 20
10 - 12 = -2	10 + 9 = 19	20 - 12 = 8	20 + 8 = 28	10 + 7 = 17
10 + 16 = 26	10 - 10 = 0	20 - 10 = 10	10 - 8 = 2	20 - 9 = 11
10 + 1 = 11	10 + 8 = 18	10 + 13 = 23	20 - 9 = 11	10 - 1 = 9
20 - 9 = 11	20 - 9 = 11	10 - 3 = 7	10 + 10 = 20	20 - 6 = 14

DAY 92 — ADDING AND SUBTRACTING 0-20

SCORE: | TIME: :

13 + 1 = 14	1 + 10 = 11	18 - 6 = 12	19 - 3 = 16	12 - 9 = 3
15 - 13 = 2	13 + 12 = 25	6 - 5 = 1	12 - 1 = 11	4 + 4 = 8
2 - 1 = 1	15 + 6 = 21	19 - 7 = 12	2 + 12 = 14	4 + 0 = 4
11 + 15 = 26	17 - 7 = 10	17 - 8 = 9	14 - 14 = 0	17 + 7 = 24
18 - 4 = 14	6 - 5 = 1	2 + 3 = 5	17 - 15 = 2	4 + 7 = 11
5 - 0 = 5	18 + 10 = 28	8 - 4 = 4	4 + 7 = 11	10 + 13 = 23
7 + 17 = 24	19 - 5 = 14	1 + 4 = 5	6 + 9 = 15	19 - 9 = 10
11 + 1 = 12	9 - 6 = 3	5 + 15 = 20	16 - 8 = 8	12 + 7 = 19

ANSWER DAY 93-96

DAY 93
ADDING AND SUBTRACTING 0-20

SCORE: | TIME: :

3 + 3 6	5 + 7 12	15 + 7 22	16 + 2 18	16 - 1 15
2 - 2 0	18 + 12 30	7 + 16 23	12 - 3 9	12 - 6 6
7 - 3 4	5 - 5 0	4 - 2 2	8 - 5 3	17 + 7 24
1 + 11 12	0 + 10 10	14 - 9 5	16 - 12 4	3 + 16 19
4 + 17 21	8 - 6 2	14 - 1 13	6 + 14 20	16 - 2 14
16 - 11 5	17 - 2 15	18 - 0 18	12 - 1 11	10 + 18 28
5 - 2 3	19 - 17 2	1 + 5 6	14 - 1 13	8 - 6 2
8 - 0 8	7 + 1 8	7 + 2 9	17 - 7 10	7 + 4 11

DAY 94
ADDING AND SUBTRACTING 0-20

SCORE: | TIME: :

20 - 12 8	20 - 14 6	10 + 2 12	10 - 1 9	10 - 6 4
10 - 2 8	10 + 0 10	10 + 18 28	20 - 4 16	10 - 7 3
20 + 11 31	20 - 7 13	20 - 9 11	10 - 12 -2	20 - 2 18
10 + 13 23	10 + 4 14	20 + 2 22	10 - 10 0	10 + 5 15
10 - 1 9	10 - 3 7	20 - 2 18	20 - 4 16	20 - 10 10
10 + 5 15	10 - 4 6	20 - 4 16	20 - 13 7	20 - 12 8
20 - 4 16	10 - 6 4	10 + 18 28	10 + 8 18	10 + 5 15
20 - 3 17	10 + 13 23	10 - 3 7	20 - 4 16	10 - 6 4

DAY 95
ADDING AND SUBTRACTING 0-20

SCORE: | TIME: :

17 + 14 31	17 - 10 7	12 + 8 20	15 + 11 26
8 - 3 5	15 - 7 8	17 - 2 15	2 + 3 5
8 - 7 1	2 + 6 8	3 + 1 4	5 + 18 23
11 - 1 10	18 - 1 17	7 + 11 18	19 - 14 5
10 - 6 4	9 + 5 14	19 - 9 10	6 - 1 5
16 - 9 7	12 - 8 4	1 + 5 6	7 + 5 12

DAY 96
ADDING AND SUBTRACTING 0-20

SCORE: | TIME: :

13 - 8 5	12 - 7 5	18 - 3 15	7 + 5 12
6 + 16 22	8 - 8 0	11 + 9 20	1 + 12 13
14 - 8 6	17 - 12 5	6 - 1 5	16 - 7 9
13 - 7 6	4 + 18 22	19 + 7 26	14 - 0 14
11 + 1 12	10 - 6 4	10 - 0 10	6 + 9 15
11 + 1 12	3 + 17 20	17 - 6 11	19 - 12 7

ANSWER DAY 97-100

DAY 97
ADDING AND SUBTRACTING 0-20

SCORE: | TIME: :

10 - 5 = 5	3 + 18 = 21	17 - 4 = 13	7 - 1 = 6
13 - 11 = 2	18 - 12 = 6	8 + 0 = 8	18 - 14 = 4
6 - 2 = 4	12 - 2 = 10	16 + 4 = 20	7 - 4 = 3
18 - 2 = 16	15 - 5 = 10	10 - 9 = 1	17 + 5 = 22
18 + 7 = 25	0 - 0 = 0	8 + 5 = 13	8 + 16 = 24
15 + 9 = 24	10 - 0 = 10	15 + 7 = 22	15 - 5 = 10

DAY 98
ADDING AND SUBTRACTING 0-20

SCORE: | TIME: :

12 - 3 = 9	15 - 9 = 6	7 + 12 = 19	10 + 7 = 17
7 + 7 = 14	4 + 1 = 5	13 - 10 = 3	12 - 2 = 10
8 + 2 = 10	11 + 5 = 16	6 + 5 = 11	11 - 2 = 9
13 + 9 = 22	1 + 17 = 18	16 - 13 = 3	16 - 7 = 9
16 - 6 = 10	9 + 8 = 17	16 - 14 = 2	13 - 5 = 8
5 + 7 = 12	18 - 1 = 17	6 + 2 = 8	16 + 9 = 25

DAY 99
ADDING AND SUBTRACTING 0-20

SCORE: | TIME: :

6 + 9 = 15	7 + 0 = 7	11 - 1 = 10	9 + 16 = 25
10 - 1 = 9	1 + 7 = 8	12 + 1 = 13	17 - 12 = 5
7 - 1 = 6	17 - 1 = 16	11 + 1 = 12	5 + 11 = 16
15 - 10 = 5	4 + 5 = 9	16 + 18 = 34	18 + 8 = 26
11 - 7 = 4	16 + 2 = 18	13 + 12 = 25	1 + 14 = 15
17 + 3 = 20	15 - 5 = 10	18 - 6 = 12	12 - 12 = 0

DAY 100
ADDING AND SUBTRACTING 0-20

SCORE: | TIME: :

14 - 14 = 0	14 + 15 = 29	18 - 3 = 15	10 - 8 = 2
6 + 17 = 23	12 + 13 = 25	11 + 9 = 20	7 - 1 = 6
9 - 4 = 5	16 - 5 = 11	8 - 4 = 4	2 - 1 = 1
9 - 0 = 9	14 + 3 = 17	15 - 2 = 13	10 - 8 = 2
18 - 8 = 10	17 - 9 = 8	19 - 6 = 13	6 - 6 = 0
6 + 10 = 16	3 + 7 = 10	15 + 16 = 31	10 + 3 = 13

Made in United States
North Haven, CT
05 November 2026

21777132R00072